SPACE COLONIZATION: TECHNOLOGY AND THE LIBERAL ARTS

AIP
CONFERENCE
PROCEEDINGS 148

RITA G. LERNER
SERIES EDITOR

SPACE COLONIZATION: TECHNOLOGY AND THE LIBERAL ARTS

GENEVA, NY 1985

EDITORS:

CHARLES H. HOLBROW
COLGATE UNIVERSITY

ALLAN M. RUSSELL
HOBART AND WILLIAM SMITH COLLEGES

GORDON F. SUTTON
UNIVERSITY OF MASSACHUSETTS AT AMHERST

AMERICAN INSTITUTE OF PHYSICS   NEW YORK 1986

L.C. Catalog Card No. 86-71675
ISBN 0-88318-347-1
DOE CONF-8510328

Printed in the United States of America

CONTENTS

CHAPTER 3: TECHNOLOGY AND SPACE COLONIZATION

INTRODUCTION

On October 11 and 12, 1985 a conference, "Space Colonization: Technology and the Liberal Arts ," sponsored by Colgate University and Hobart and William Smith Colleges, was held in Geneva, New York. This volume contains the proceedings of that conference. A remarkable variety of papers was presented and several significant themes emerged which are described below. First, however, we discuss what led to holding a conference on this unusual combination of subjects.

SPACE COLONIZATION

Throughout history the dream of living off the Planet Earth has recurred with varying degrees of realism and fantasy. In this century the dream has come close to being reality. Current awareness of the possibility has been greatly stimulated by the efforts of Gerard K. O'Neill, a Princeton University professor of physics. In 1974, after several years of consideration, O'Neill published an article in *Physics Today* entitled "The Colonization of Space" that kindled public interest as well as the interest of academics and scientific and engineering professionals. His books, papers, and technical contributions have continued to stimulate interest.

In the summer of 1975, as a direct result of O'Neill's efforts, the American Society for Engineering Education, the National Aeronautics and Space Administration, and Stanford University sponsored a ten-week long study of the feasibility of establishing large communities in orbit somewhere in the vicinity of the Earth or Moon. Some twenty faculty and ten students and visitors worked for ten weeks at Stanford University and at NASA's nearby Ames Research Center to design a system that would support such a space city. Their disciplines included biology, economics, sociology, anthropology, chemistry, computer science, physics, and chemical, civil, mechanical and industrial engineering. A comprehensive and technologically consistent plan for an orbiting city was devised; it is described in Space Settlements: A Design Study, NASA-SP 413, edited by Richard D. Johnson and Charles Holbrow, (Government Printing Office, Washington, DC, 1977).

PEDAGOGY AND SPACE COLONIES

Many of the participants of the 1975 Summer Study saw clearly that the idea of space colonies had exciting potential as a teaching device. This usefulness has been apparent to other teachers too, and in the past decade faculty have incorporated the idea of space colonization into courses on architecture, philosophy, physics, sociology, anthropology, political science, and economics. The following papers also show that consideration of space colonies continues to lead to interesting and creative ideas in these and other disciplines.

Because of the growing interest in using space colonies to motivate students and faculty to look with fresh eyes on their particular academic material, we decided to organize this conference. We felt it was time for faculty from diverse disciplines to share what they had done and were planning to do.

Current interest in interdisciplinary studies also made a conference timely. A space colony is a microcosm of society. To discuss the establishment of even a small community in space involves a broad spectrum of academic disciplines. Combined with its obvious dependence on a variety of technologies, the space colony is an exceptional subject for interdisciplinary study. Just a brief examination brings home to the student the complexity and interrelatedness of a human community. He or she quickly sees the need to consider social organizations from many different perspectives.

Human existence is critically dependent upon technology; this dependence is another important reason to introduce space colonies into the curriculum. We live in a world shaped by technology, but as a society we are blindly ignorant of the nature of technological dependence.

The technological literacy of many people who substantially influence decisions and public opinion is very poor. Misunderstanding and ignorance of important engineering concepts of risk and benefit, of trade-offs, of systems design, have led the public into unrealistic expectations of technology. Obviously it is an important responsibility of our educational institutions to lessen this technological illiteracy. Equally clearly, it is essential that as a society we educate decision makers to be capable of making informed judgments about complex technological problems so that we may continue to develop the technologies that will permit our society to progress in ways in accord with human aspirations.

It is also desirable to place technology in a larger social and political framework. Harvey Brooks reminds us that technology is more than hardware. It is "...socio-technical rather than technical, and a technology must include the managerial and social supporting systems necessary to apply it on a significant scale." These people who work these systems need to comprehend their role. We all need a broad view of what we are about. Otherwise, we work our ways into technological dead-ends. Without a clear, broad, social vision we generate SSTs only to have them rejected by society; we produce nuclear power but for a society unwilling to have it; we go to the Moon, but then have no longer-term, societal goals to which this technological triumph contributes.

Without this broader, liberal-arts education people are helpless to make intelligent decisions about technology. How are we to understand the Challenger tragedy? Is it to be seen as a profound failure of our technology? Is it to have been expected? Or what of the Chernobyl disaster? Is it planet threatening? And if it is, what are our alternatives? Are we, as Henry Adams wrote in 1852, "slaves of our power houses?" Very similar questions can be raised in connection with space colonies. An advantage is that they can be raised in a context less emotionally and politically charged than that which surrounds the events of the evening news.

The recognition of a confluence of

-a need for technological literacy
-the importance of setting technological goals in a larger context
-the need to educate students to examine complex enterprises from the perspectives of many disciplines
-the desirability of tapping student interest and enthusiasm for space in order to stimulate thought about other issues of broad human concern

led us to propose a conference on "Space Colonization: Technology and the Liberal Arts," and we issued a call for papers to describe how these and related questions are or can be introduced into the liberal-arts curriculum.

Response was gratifying. We were particularly pleased with the variety. The humanities, the social sciences, and the physical sciences are all represented here. It is interesting to note that three of the innovative courses described here were developed out of a collaboration of physicists and sociologists or philosophers. The variety of perspectives was further broadened by the participation of an appreciable number of people who had worked on the 1975 Summer Study, including several non-academics — a physician, an author, and two engineers from industry. Exchanges across disciplines and professions were informative and stimulating throughout the conference.

SPACE COLONIES: HEURISTIC DEVICE OR ACTION PROGRAM?

The conference papers are presented in three chapters. Chapter 1 consists of descriptions of actual uses of space colonies in the liberal-arts curriculum. Chapter 2 is a collection of proposed uses and insights stemming from thoughtful consideration of space colonization. Chapter 3 contains a history of old and proposals for new technology to make space colonization feasible. The chapter contains a symposium in which conference participants who had also worked in the 1975 Summer Study looked at the prospects for space colonization.

Space Colonies as a Heuristic Device

The papers in Chapter 1 describe uses of the space colony as a model society, a *gedanken* world, by which to illustrate ideas and concepts of philosophy, physics, architecture, sociology, and economics. The philosopher Gonzalo Munévar, University of Nebraska, Omaha, offers an examination of the philosophical basis for space habitation as it is presented in a philosophy course that he and a physicist colleague have developed around the theme of space colonization and the basic questions it raises. David Van Blerkom, University of Massachusetts, Amherst, presents an ingenious discussion of the behavior of a baseball in a rotating space colony. He has used this discussion to interest and educate liberal-arts students in physics. Patrick Hill, California Polytechnic State University, San Luis Obispo, describes ten years of using the ideas of space colonization in his teaching of architecture. William MacDaniel, Niagara University, tells of the problems, successes, and prospects of his sociology course "Living in Extraterrestrial Space." Mona Cutolo, a sociologist, and Denis Miranda, a physicist, describe the comprehensive course, "Space: A New Frontier," that they have developed as part of Marymount Manhattan College's attempt to develop technological literacy among liberal-arts students. Finally, Martin Giesbrecht gives a brief analysis of how his familiarity with space colonization has changed his teaching of economics.

Chapter 2 consists of proposed and planned curricular uses of space colonization plus some interesting insights that the subject generates. Barry Turner, University of Exeter, U.K., presents a thoughtful description of the sociology course he has begun to teach. This paper should be an exceptionally useful resource for other sociologists planning to use space colonization in their courses. Thomas L. Melchionne and Steven L. Rosen of Rutgers University describe how the idea of space colonization can serve the teaching of anthropology. Fay Terris Friedman, D'Youville College, outlines possible applications to the teaching of psychology. Marvin Israel and T. Scott Smith, Dickinson College, propose a different use of space colonization as a teaching device. They ask how closely is the essential "human nature" of *Homo sapiens* tied to our connection with this Planet Earth. They ask if separation from the planet will fundamentally change the nature of humanity. And they ask would such a change be acceptable. Steven Lee and Scott Brophy of Hobart and William Smith Colleges use the theme of space colonization to raise basic philosophic questions about when, whether, and how technological options should be pursued. They argue that massive public investment in space colonies would be immoral.

Chapter 2 contains two papers that grapple with interesting issues of space colonization outside an explicit curricular framework. Barbara Eckman, University of Pennsylvania, explores why the idea of space colonization has such strong appeal to so many people. She does this from the perspective of the humanities and reveals new and thought-provoking aspects of space colonization. Joel Scheraga, Princeton University, uses the prospective movement of people into space to motivate his concern for how the resources of space will be allocated. He argues that we need to develop sensible modes of resource allocation and that space will be most efficiently developed and used if property rights are assigned to those who will exploit it. He proposes "homesteading" of space resources as a way to vest such rights. He makes a convincing case that these matters should be under consideration right now.

Chapter 3 contains two papers and the transcript of a panel discussion. The first of the papers was delivered as a public lecture by Arthur Kantrowitz, Dartmouth College. Professor Kantrowitz was the featured speaker of the conference. He calls for technological optimism and suggests why and how we should pursue technologies that will foster movement into space. In the second paper Thomas Heppenheimer, author, describes his personal involvement in and recollections of the events of the 1970's that led to an appreciable public awareness of space colonization. The panel, composed as it was of participants in the 1975 Summer Study, also expresses recollections of some of those events. The members of the panel assess the progress to date and the likelihood of further progress toward space colonization.

Space Colonization as a Program of Action

A concern for space colonization as a program of action is explicit in the papers of Chapter 3: Should it be done? This concern was present implicitly or explicitly in many of the other papers as well. Consequently, mingled with exploring space colonization's uses as a heuristic device were attempts to grapple with its desirability or feasibility. Of course, this too has its heuristic uses, and several of the proposed courses were built around answering the question. Munevar's paper in Chapter 1, a balanced philosophical analysis of this question, answers "yes." Kantrowitz in Chapter 3 provides a technological justification. Lee and Brophy, in Chapter 2, argue that it would be immoral, a waste of public monies; Israel and Smith raise serious questions about the desirability of a series of acts that could fundamentally alter the very essence of human nature. In the panel discussion deep philosophical divisions appeared between those who view technology and its potential with enthusiasm and those who view it with suspicion. The familiar debate between environmentalists and developers emerged in a slightly different guise during the panel discussion.

In this connection Barbara Eckman points out in her paper in Chapter 2 that whether the reasons for going into space are convincing or not, there exists in many people a powerful yearning to go. From a Jungian point of view she explores the non-rational motives for going: Why do people really want to leave the planet? What is the source of what Winkler called "the insane, magical feeling" associated with the idea of space colonization? Her answer is consistent with the point she made later during the panel discussion that human movement into space can be harmonious in spirit, rather than fiercely exploitative, without denying the practical difficulties and the daunting challenges.

Some Conclusions

The papers present some important guidance for the teacher who proposes to use space colonization as a theme by which to motivate discussion of the principle ideas of any of a variety of disciplines. Two warnings are especially worth noting. MacDaniel warns of the "over-advocacy trap." Do not assume that all students will be aware of or enthusiastic about space colonization. And several teachers report the need for caution in asking students to deal comprehensively with the subject. Don't ask students to make sweeping and grandiose analyses. It is important to break the topic into accessible parts in order to avoid overwhelming the student.

It is also clear that the teacher should understand to what extent he or she is an advocate or opponent of space colonization. That position should be explicit, and it should be informed. The papers presented here should serve to inform the reader of the variety and thoughtfulness of these positions as well as to stimulate further use of the ideas in the liberal-arts curriculum.

AN OMISSION AND SOME USEFUL INFORMATION

One especially interesting presentation is not included in these proceedings because it depended heavily on visual aids that can not be reproduced here. Harold Jebens, Marquip Corporation, gave a memorable slide talk describing the basic proposal for space colonization that came out of the 1975 Summer Study. Because these proceedings frequently allude to the 1975 Summer-Study design, we summarize Dr. Jebens' talk here.

The overall system included three main parts: the Earth, a space colony located in orbit equidistant from Earth and Moon at the Lagrangian libration point L5, and a mining facility on the Moon. After an initial construction module from Earth has been put in place at L5, sintered blocks of ores would be launched from the Moon by electro-magnetic catapult (mass driver). These would be recaptured in space and transported to L5 where they would be refined using solar energy. The colony would then be built from the refined metals. The colony would sustain itself economically by manufacturing solar power generating plants that would be placed in geostationary orbit where they would generate electrical power to be sold to Earth. Concentrated microwave energy would be beamed to Earth, converted into electricity and deliverd over the Earth's power distribution network. The ideas for using power satellites drew heavily on the work of Peter Glaser of the Arthur D. Little Corporation.

The orbiting colony of 10,000 people was designed in the form of a torus, a giant bicycle wheel. This design is referred to as the "Stanford Torus." The radius of the wheel would be about 900 m and the diameter of the "tire" of the wheel would be 130 m. People would live inside the "tire" and pseudo-gravity would be provided by rotation of the wheel about its hub. Extensive technical details are given in *Space Settlements: A Design Study*. The overall design of the space-colony system drew much from the work of Gerard K. O'Neill who was the Technical Director of the Summer Study. Dr. Jebens described this design in a lively talk using the excellent graphics developed by NASA artists.

ACKNOWLEDGMENTS

We are grateful to Hobart and William Smith Colleges and to Colgate University for their sponsorship and support of this conference, the latter using funds from a Sloan Foundation grant for the New Liberal Arts. Modest contributions were also received from the Faculty of Social and Behavioral Sciences and the Department of Sociology within that Faculty and from the Massachusetts Institute for Social and Economic Research, all at the University of Massachusetts, Amherst. We thank these institutions for their contributions.

Charles H. Holbrow,
Colgate University
Allan M. Russell,
Hobart & William Smith
Colleges
Gordon F. Sutton,
University of Massachusetts
at Amherst

SPACE COLONIES: TECHNOLOGY AND THE LIBERAL ARTS

List of Participants

Steve Brody
Intermetrics, Inc.
Houston, TX

Mona Cutolo
Department of Sociology
Marymount Manhattan College
New York, NY

Fay T. Friedman
D'Youville College
Buffalo, NY

Thomas A. Heppenheimer
Center for Space Science
Fountain Valley, CA

Charles H. Holbrow
Department of Physics and Astronomy
Colgate University
Hamilton, NY

Harold J. Jebens
Marquip, Inc.
Phillips, WI

Steven Lee
Department of Philosophy
Hobart and William Smith Colleges
Geneva, NY

Thomas L. Melchionne
Department of Anthropology
Rutgers University
New Brunswick, NJ

Gonzalo Munévar
Department of Philosophy
   and Religion
University of Nebraska
Omaha, NB

Scott Brophy
Department of Philosophy
Hobart and William Smith Colleges
Geneva, NY

Barbara Eckman
Department of Religious Studies
University of Pennsylvania
Philadelphia, PA

Martin G. Giesbrecht
Economics Associates
Wilmington, OH

Patrick D. Hill
Architecture and
   Environmental Design
California Polytechnic
   State University
San Luis Obispo, CA

Marvin Israel
Department of Sociology
Dickinson College
Carlisle, PA

Arthur R. Kantrowitz
Thayer School of Engineering
Dartmouth College
Hanover, NH

William E. MacDaniel
Department of Sociology
Niagara University
Niagara University, NY

Denis Miranda
Department of Physics
Marymount Manhattan College
New York, NY

Ruth Reinsel
Department of Psychology
The City University of New York
New York, NY

Rowland Richards, Jr.
School of Engineering
State University of New York
Buffalo, NY

Allan M. Russell
Department of Physics
Hobart and William Smith Colleges
Geneva, NY

T. Scott Smith
Department of Physics
Dickinson College
Carlisle, PA

Barry A. Turner
Dean of Social Sciences
University of Exeter
Exeter, England

Steven L. Rosen
Department of Anthropology
Rutgers University
New Brunswick, NJ

Joel D. Scheraga
Department of Economics
Princeton University
Princeton, NJ

Gordon F. Sutton
Department of Sociology
University of Massachusetts
Amherst, MA

David Van Blerkom
Department of Physics
University of Massachusetts
Amherst, MA

Lawrence H. Winkler
Winnipeg, Manitoba
Canada

CHAPTER 1

SPACE COLONIZATION
IN THE CLASSROOM

This chapter presents six papers describing applications and uses of the ideas of space colonization in the teaching of philosophy, physics, architecture, sociology, general education in technology and society, and economics.

SPACE COLONIES AND THE PHILOSOPHY OF SPACE EXPLORATION

Gonzalo Munévar
Department of Philosophy
University of Nebraska at Omaha

ABSTRACT

Many space enthusiasts believe that the possibilities offered by space colonies clinch the case in favor of space exploration. Such possibilities, however, cannot by themselves surmount the central social and ideological objections against space exploration. Moreover, to justify the process by which we can determine whether space colonies are a good idea requires that we meet those objections first. This task is often attempted by pointing to the many unintended good results of previous exploration (the serendipity of science) and then extrapolating to the future. But social and ideological critics need not be impressed by a purely historical case for serendipity. Fortunately, a philosophical analysis of scientific exploration reveals that serendipity is an essential aspect of it. This result provides a justification for exploring space. And in light of that justification, we can begin to evaluate the proposals for space colonies.

INTRODUCTION

Space colonies present many opportunities for thinking about traditional philosophical issues. Discussion of the social and political structures of space colonies, for example, bring to mind Plato±s Republic and J.S. Mill's On Liberty. But space colonies serve an even more important role in thinking about new philosophical issues -- issues connected with the nature and justification of space exploration. It is about those issues that I wish to speak in this paper.

0094-243X/86/1480002-11$3.00 Copyright 1986 American Institute of Physics

SHOULD WE GO? ARGUMENTS PRO AND CON

At the University of Nebraska at Omaha, a colleague from the physics department, John Kasher, and I have developed an interdisciplinary course entitled "The Philosophy of Space Exploration," taught at the upper division and graduate levels. For this course students can receive credit in either physics or philosophy. The course is enormously successful, but I will say no more about that here. I wish instead to give some idea of the content of the course; more specifically, I wish to sketch the role that the proposal for space colonies plays in one of the main philosophical disputes that we consider during the semester. That dispute concerns the justification of space exploration and is the central topic of a book that I am writing, The Dimming of Starlight.[1] Although it is not really a textbook, an early draft of this book has recently served as such for our course at the University of Nebraska at Omaha and for a graduate seminar on the same subject that I taught at Stanford University a year ago.

Con

In this book, and in the course, I point out that many people seem to think that the possibility of constructing space colonies clinches the case in favor of space exploration. There are basically two main kinds of objections that must be overcome to justify space exploration: social and ideological. Social objections are to the effect that space exploration takes money, talent, and effort away from more pressing human needs (such as combating poverty and hunger). Ideological objections hold that space exploration is an unwise activity, a big extension of the science and technology that, coupled with the mentality of growth, have done so much to destroy our environment, deplete our resources, and bring our planet to the brink of disaster. The problem is with the mentality that leads us to interfere with and exploit nature instead of trying to live in harmony with it.

Pro

In reply to these objections, space enthusiasts tend to list the many benefits we derive from the space program: weather satellites save lives and crops, communication satellites bring about an economic expansion, land satellites discover resources and help us monitor the environment, and space technology spins off valuable products into our lives. Space exploration thus contributes greatly to the reduction of human misery, the improvement of human life, and the preservation of the environment. That is the standard case in favor of space exploration.

Moreover, we are just beginning to move into space. The low gravitation and the vacuum of space offer many industrial and technological advantages. All these industrial and technological advantages, however, merely point to the way in which space exploration may play a major part in solving some of the most urgent problems of the Earth. Our world faces at once an increasing demand for energy and a dwindling of resources. In trying to obtain more energy we use up even more resources and, to make matters worse, produce greater amounts of pollution, which in turn affects some of our other resources, to say nothing of our health and general well-being. When we do obtain more energy through fossil fuels, which are the usual source of industrial energy, we release ever greater amounts of CO_2 into the atmosphere. If the rate at which CO_2 is released continues to increase, some observers fear that the resulting greenhouse effect might raise the temperature of the planet enough to melt the polar caps. And if the polar caps melt, the weather will change and the oceans will rise. If the oceans rise, many of the great cities of the world will be destroyed, and large areas now inhabited by millions of humans will be flooded out of existence.[2]

To forestall these dire consequences, space enthusiasts have made proposals that range from the building of solar power satellites to the mining of the Moon, the asteroids, and eventually other planets. A solar power satellite can collect sunlight, transform it into electrical energy, and beam that energy down to Earth. With this means of procuring energy we can prevent the social and economic dislocation caused by scarcity, while avoiding the environmental catastrophe that a vast increase in the use of fossil fuels courts. In space the sunlight is plentiful and likely to last for billions of years; solar power satellites release no CO_2; and environmental studies indicate that beaming this energy would be less harmful to plant and animal life than the existing alternatives. One gigantic solar power satellite the size of the island of Manhattan would provide as much power as ten nuclear power plants, without the attendant risks of radioactive leaks and meltdowns.[3] As an added bonus, other enthusiasts have suggested moving some of the most polluting industries to space. The promise of space exploration is then very enticing: abundant energy and a safer, cleaner environment.

Critics of these proposals have argued that the extraction of the enormous quantity of materials required to build such structures would cause major environmental headaches, while the many thousands of flights by giant rockets to haul the materials into orbit may damage the atmosphere and are certain to cost far too much. But this criticism is misleading. The physcist Gerard O'Neill, for example, has said all along that most of the required materials (e.g., aluminum, oxygen, and silicon) can be rather easily extracted from the Moon, placed in lunar orbit and processed there.[4] The gravitational pull of the Moon is only one-sixth that of Earth, and thus the materials can be shot into lunar orbit by what O'Neill calls "mass drivers," at great savings of energy and money. This project would be the beginning of the eventual colonization of the solar system, for no insurmountable technological barriers would keep us then from the abundant resources available in the asteroids, nor from building large habitats in space. To paraphrase O'Neill, the closing of the Earthly frontiers would be compensated by the opening of the high frontier to the hopes and needs of humankind. And that presumably clinches the case for space exploration.

ANALYSIS OF THE ARGUMENTS

At this point we do not need to discuss the subtleties of every possible reply by the critics, since the standard case for exploration encounters serious obstacles in the core of the original objections. Let us take the social critics first. In spite of these long lists of actual and potential benefits, it soon becomes apparent that the standard case does not go far enough. For many important space activities do not have the obvious beneficial consequences of weather and communication satellites. Where is the obvious pay-off of a probe of Jupiter or Titan, of landing a vehicle on Mars, of scooping a bit of Halley's comet? Few accomplishments of space exploration rank as high as the discoveries made with telescopes in orbit. But how is that information from space astronomy going to put food in children's mouths or a roof over their heads?

The Argument of Practical Benefits

In emphasizing the practicality of space technology, the standard case makes intellectual orphans of those very things that bring to exploration an air of mystery and excitement. What it leaves out is the heart of space exploration: those activities motivated by our sense of adventure, by our urge to explore, by the need to satisfy our curiosity. That is why this justification along practical lines does not go far enough. It fails because it excludes those aspects of the enterprise that ignite the imagination and stir the soul about the conquest of the cosmos.

Arguing that scientific knowledge has value in itself only brings us back to the original debate. Is scientific knowledge more valuable than achieving this or that other social aim? We have not answered that question yet. Nor is the cause of exploration advanced by pointing out that since only a portion of the space budget is allocated to science (while most of it presumably goes for the more obviously practical activities), and since the space budget is not so large to begin with, taking the money away from the heart of space exploration is not going to solve the social problems anyway. Thus, since at least some of the social critics will admit that scientific knowledge is valuable, they should be prepared to admit also that such a small fraction of the GNP is well invested in the search for knowledge. One important reason many social critics would not accept this reply is that even though the fractions are small the actual sums are large. The price tag for one proposed space telescope alone is around $1.5 billion. That sum would not solve all the social problems of the world, but a lot of good might be done by its judicious investment. And the question once again becomes whether we should do all that social good or search for knowledge instead. The more ambitious the plans of the space enthusiasts, and thus the greater the corresponding budgets, the more pressing this question becomes.

Space enthusiasts might respond, however, that a crucial aspect of the case has not been presented adequately. All those benefits listed are the results of having yielded earlier to the call of the heavens. When we first explored, we did not know for certain that so many good things would repay our efforts; very often we had no inkling. The pursuit of scientific exploration pays because of the serendipity of science; that is, because of the unintended benefits that science yields. And now the space enthusiasts want us to share their faith in the future of exploration, to believe with them in the continuous flow of treasure from our space ships, even when they cannot say what that treasure will be.

Unfortunately, the critics may doubt that prior performance is enough warrant for that faith. Having gotten water out of a well before does not guarantee an inexhaustible supply. Even space activities near Earth, which are often beneficial because of the vantage point they provide, are beginning to experience problems of saturation. Geosynchronous orbit, for example, is becoming crowded with communication satellites that are beginning to interfere with each other. And space debris -- mostly from the break-up of rockets -- is becoming a hazard to operations in lower orbits.[5] Advances in technology will probably solve these problems, but still we can see that linear growth of benefits is not automatic.

The evidence for serendipity becomes more tenuous the farther we go away from Earth. No one has yet provided a link between a probe of Jupiter's atmosphere and the lot of those who breathe the atmosphere of Earth. Even then history alone would not suffice, for the critics may still suspect that the serendipity of space explora- tion has been a fortunate accident. Moreover, that history is often dominated by confirming anecdotes (for example, by remarks about how the research on electromagnetism by the 19th Century Scottish physicist Clerk Maxwell made possible television and computers, two inventions which Maxwell himself could not have foreseen). But nothing is ever said about the overwhelming majority of the research carried out at the same time. Does all of science yield practical benefits, or only some of it? The critics may well suspect, say, that only exceptional science is gifted with serendipity. How can space enthusiasts show that the critics are wrong on this matter? Or else how can they tell beforehand that the research they propose will prove to be exceptional? As it is, they would have great difficulty even showing that exceptional science guarantees serendipity. What those enthusiasts need is an account of science in which unintended good results are seen as unavoidable. Until they have it, their efforts at justification along practical lines will expose the heart of space exploration to the narrow-minded whims of cost-benefit analysis. That is hardly the stuff dreams are made of.

Other social critics may well believe that planetary exploration and space astronomy drive technology to some degree. But they may still question whether that drive is substantial enough to justify setting aside the satisfaction of other human needs.[6] Nevertheless, as far as all the social critics are concerned, there is an even more serious objection: if spin-offs are so valuable, does it not make more sense to spend the money directly in the relevant fields?

Moral Limits

The ideological critics should be even less impressed by a justification that involves technological and economic growth. Indeed, they see the alleged benefits as causes for concern. To many of these critics, and especially to some who are influenced by the environmentalist movement, the very idea of space exploration is not only unwise but immoral. This judgment they pass with particular harshness on some of the grandiose proposals for solving the most pressing Earth problems by going into space. According to Wendell Berry, for example, the lesson that we should learn from the closing of the earthly frontiers "calls for an authentic series of changes in the human character and community that, if made, will afford us the spiritual resources to live both within our material means and with each other."[7]

Space exploration, on the other hand, tries to outflank the lesson entirely. The space enthusiast -- and here Berry has Gerard O'Neill in mind -- ignores what is essentially a moral problem (i.e., the changing of human character and community) and offers technological solutions instead. The morality of the space enthusiast is thus both shallow and gullible, for what is offered is "a solution to moral problems that contemplates no moral change."[8] Space exploration, to someone like Berry, could only be a "desperate attempt to revitalize the thug morality of the technological specialist, by which we blandly assume that we must do anything whatever that we can do."[9] According to another critic, Dennis Meadows, "What is needed to solve these problems on earth is different values and institutions -- a better attitude towards equity, a loss of the growth ethic. . . . I would rather work at the problems here."[10] At first sight the position held by Berry seems question-begging. The closing of the human frontiers presents a moral problem to which only moral solutions are applicable. Gerard O'Neill and other space enthusiasts ignore the moral problem. Thus they are not only doomed to failure but are also immoral (not just mistaken or unperceptive). But what O'Neill and the others question is precisely whether there is a moral question. They do so by advancing a variety of arguments against the claim that the frontiers have closed. That claim is what is at issue, and thus Berry's objection begs the question.[11]

Technology vs. Nature

But perhaps there is a more sympathetic reading of Berry's position. What he may have in mind is that the experience of the closing of the earthly frontiers (or even coming so near to that closing) is enough to show that the Western society's approach to nature is inherently unwise, and thus that its extension through space exploration is destined to fail. On what grounds should we trust O'Neill's grandiose plans for gigantic solar power satellites, let alone those for artificial worlds (his space colonies)? Surely projects of such magnitude cannot be made plausible merely by theoretical expositions in which everything comes out right. How can we be assured that no essential detail has been left out? Indeed, there are many reasons for suspecting that these grandiose plans overlook some crucial obstacles. For example, it is assumed that a whole ecological system can be built from scratch without great difficulty. Ecologically minded opponents of exploration think that this is one more instance of the arrogance of physical scientists in their approach to nature. Proponents of space colonies often argue that the exploitation of the resources of the solar system can be rather painless thanks to the use of automata, and extremely simple and inexpensive if we were to use von Neumann machines (self-replicating automata). But there are serious reasons for doubting that such machines are technologically possible.[12]

The urgency of the situation, as these ideological critics perceive it, makes unwarranted our engaging in any more technological detours. Western society's approach has brought the world to the edge of crisis by marrying technology to the mentality of growth. This ideological criticism touches the heart of space exploration insofar as science is supposed to provide the promissory note that underwrites that marriage in the first place. Indeed the satisfaction of scientific curiosity -- at least where "big" science is concerned -- may be seen as a disturbance, as an interference with nature. The emphasis on beneficial results only blurs the overall picture. So the ideological suggestion is that Western science is on the whole unwise because its practice is somehow unnatural (against nature, at any rate). And surely, the suggestion goes on, practices which are not in harmony with nature are bound to fail. Space exploration remains, then, a distraction at a time of crisis -- the siren voice that calls us from the cosmos sings the tune of our doom.

Humankind Needs Science; Science Needs Space

A philosophically and rhetorically satisfying case for space exploration needs two things. First it needs to show that serendipity is a natural byproduct of science. And in showing this it has to say something about the wisdom of the whole enterprise. This double task can be accomplished, I believe, by paying attention to the nature of science, that is, by treating the problem of justification, at least in part, as a problem in the epistemology of science.

It is normally thought that facts provide the warrant of science, by either proving or disproving theories in some way. Nevertheless, I think that the emphasis on facts is misleading. Science goes beyond the collection of facts in a fundamental way, and this is a key to the transformation of the debate. The supporters of exploration may still come out ahead.

As sensible as the conventional view on the nature of science may sound, it is rendered unacceptable by recent developments in the philosophy of science. One lesson we learn from contemporary thinkers like Paul Feyerabend, Thomas Kuhn, Imre Lakatos, and to some degree Karl Popper, is that scientific views or theories are like spectacles through which we experience the universe.[13] We might even say that they are instruments by which we conceive of the universe. According to Kuhn, for example, scientific views (which he calls "paradigms") tell us what elements there are in the world, what relations exist between those elements, and thus what sorts of problems are meaningful in a particular science, as well as what kinds of answers are acceptable (for a scientific view determines the theoretical, mathematical, and experimental commitments of the discipline).[14] To speak of science as a pair of spectacles that permit us to see the world is to speak metaphorically, to be sure, but that metaphor is by no means farfetched. We should realize that without our scientific views we would simply be blind to many aspects of the universe. And those views give us more than pictures of the universe; they also provide means of interacting with it.

Science, then, gives us more than mere representations of the universe. For science asks questions by seeing, hearing, analyzing, probing, and touching nature at many different energies and magnitudes. Thus, when we learn to "see" the universe, we actually learn to make contact and deal with its diverse facets in many different ways.

When we place emphasis not on facts but on the essential trans-formations of science and their consequences, we gain a new point of attack. Since our world views tell us what the world is like, they also determine ultimately what opportunities we can take advantage of and what dangers we can be warned about. Thus, with changes of world view comes the realization of many new opportunities and dangers. And changes in world views are inevitable given the nature of science. For scientific theories -- the underpinnings of our world views -- are still our creations and thus imperfect. They are always in need of refinement, modification, or replacement altogether. The pressure for such changes comes from the double exposure to unusual circumstances (which force us to stretch our views) and to competing ideas (which are often developed to account for a few of those unusual circumstances, and then claim the right to extend to the entire field of the discipline). Thus, the essential feature of science is not merely the addition of a few, or even many, interest-ing facts but the transformation, perhaps radical, of our views of the world.[15]

This essential feature turns serendipity into a natural consequence of science. If science is to be science it must be challenged; it must change; it must be a dynamic enterprise. But that change is a change in conception, a transformation of our views of the universe. And once we think about the universe differently, once we have a different "communal" perception of it, we come to perceive hitherto unknown dangers, new solutions, new opportunities. Such is the cradle of serendipity.

Armed with this result, the supporter of exploration can answer the social critic and begin his reply to the ideological critic. But this approach to the controversy makes sense only if we grant a crucial assumption, namely that science and space exploration go hand in hand. This may seem obvious to a casual observer, but it has been bitterly contested over the years. Many scientists, perhaps the majority of scientists, were opposed to the Apollo program to put a man on the Moon, on the grounds that it was political showbiz and not science.[16] And just about every important field of space science has been looked down upon, at one time or another, in the most prestigious and established quarters of science.[17] Some of them still are.[18] And, if we pay attention, we may still hear rumblings that all that money should go for really fundamental research.

It is therefore crucial to determine in which sense, if any, space science can be said to be fundamental science. This task in turn requires that we pay close attention to the actual practice of space science; that is, that we discuss in some detail the goals, methods, and procedures of each of the space sciences. Only then can we begin to pass judgment on the scientific merit of comparative planetology, space astronomy and physics, and space biology. But lest the reader be weary of a task that demands unreasonably wide expertise, I should specify that to proffer this judgment we just need to see whether the general remarks made about science earlier can be extended to these branches of space science. Once again, success will depend mostly on a sensible application of philosophy of science.

This can be done, although I cannot describe the particulars here. Once it is done, we will find the necessary warrant to commit ourselves to space exploration on a large scale. At that point we might have the opportunity to build and test ever more complex structures in space. We would be able, thus, to evaluate the proposals for space colonization. But should we follow that path in our exploration of space? All such proposals, even those conceived strictly for the benefit of the mother planet, require the cutting of the umbilical cord. Earth may still be home, but our dreams begin to pull us away from it.

Some space scientists believe that we should not follow those dreams, that the only exploration worth doing involves science. But to do scientific exploration, they suggest, we are better served if we operate in space by remote control or let robots go in our stead.

At this juncture the issue of space colonies becomes very important once again. For if we can show that machines cannot replace humans in space, and I believe we can, then space colonies may have a clear sailing. The reason is that we would have already

shown that space exploration is justified, and now we would need only
determine what would be the best approach to it. And surely, in this
context, the promise inherent in the proposals for using the
resources of the solar system makes it reasonable to begin the
process of evaluation. If that process is successful, if more
complex space undertakings eventually lead to large space habitats,
and then to colonies in space, the dreamers of today will be the
founding fathers of the bountiful tomorrow. But the journey is long
and the obstacles many, and thus we need a clear vision of the road
that awaits us. For only then will our steps move us forward.

REFERENCES AND FOOTNOTES

1. Several passages in this paper are excerpts from that book. A
    report on the course (Physics 305/Philosophy 305) appears in
    Social Sciences and Space Exploration, NASA EP-192, edited by
    T.S. Cheston, C.M. Chafer, and S.B. Chafer, (Government Printing
    Office, 1984), p. 104.
2. Of course the change in weather may also be beneficial to some
    areas. A warm Siberia, for example, may become one of the
    largest gardens of the world.
3. For details see Gerard O'Neill's The High Frontier 2nd edition,
    (Anchor Press/Doubleday, New York, 1982). See also T. Heppen-
    heimer, Colonies in Space, (Stackpole Books, Harrisburg, Pa,
    1977).
4. O'Neill, op. cit.
5. Orbital Debris, NASA CP-2360.
6. Until now NASA has had a policy of bringing about technological
    breakthroughs with each new mission to explore the solar system.
    Because of budgetary constraints that policy apparently will
    change. Many future missions will depend on a recycling of
    existing technology. It seems that the scientific exploration of
    space need not drive space technology substantially anymore. The
    impact of the esoteric technology used to explore Jupiter and
    Saturn on the general technology cannot be discounted, but
    estimating that impact precisely, or even approximately, is not
    an easy matter. For a discussion of this point see Mary Holman,
    The Political Economy of the Space Program, (Pacific Books,
    1974), and Mary Holman and Theodore Suranyi-Unger, Jr., "The
    Political Economy of American Astronautics," AAS Paper No.
    80-51, March, 1980.
7. Wendell Berry in Space Colonies, edited by Stewart Brand,(Penguin
    Books, New York, 1977), p. 36.
8. Ibid.
9. Ibid. p. 37.
10. Dennis Meadows, Space Colonies, p. 40

11. And then, by all appearances, he piles abuse on top of bad argument.
12. I discuss the implausibility of von Neumann machines as tools of colonization or exploration in Chapter 8 of The Dimming of Starlight.
13. The most important references advancing this view are: Paul Fayerabend, Against Method, (NLB, 1975); Thomas S. Kuhn, The Structure of Scientific Revolutions, 2nd edition, (University of Chicago Press, 1970); Imre Lakatos, Philosophical Papers, (Cambridge University Press, 1978).
14. Kuhn, op. cit.
15. This theory of science is developed in detail in my Radical Knowledge: A Philosophical Inquiry into the Nature and Limits of Science, (Hackett, 1981 [Avebury in the U.K.]).
16. How general the feeling was in the scientific community is il- lustrated by a survey done in 1964 by Science, the journal of the American Association for the Advancement of Science. 16% of the science Ph.D.'s who responded agreed with President Kennedy's decision to go to the Moon, while an overwhelming 64% disagreed. Even today, in giving talks about the nature of space science I am often confronted with the objection that if science is our goal we should spend the money on particle physics instead.
17. For an account of the low status suffered by the planetary sciences until rather recently see Stephen G. Brush, "Planetary Science: From Underground to Underdog," Scientia, 113, 771, (1978). Brush demonstrates how the prejudice against planetary science was completely blind to the history of physics.
18. Especially space biology.

BASEBALL IN SPACE
Space as a Unifying Theme in Physics for Non-Science Majors

David Van Blerkom
Department of Physics and Astronomy,
University of Massachusetts, Amherst, MA 01002

ABSTRACT

The theme of space colonization serves to unify the topics presented in a physics course for non-science majors. Some vivid examples of the behavior of a baseball in space can bring home to students some of the odd features of the simulation of gravity by rotation. A pop fly or a pitched fastball may behave strikingly differently in a rotating space habitat than on Earth. The differences are derived by simple calculations that use only elementary physics.

INTRODUCTION

The topic of artificial space colonies figures in an introductory physics course I have been designing for students in the humanities. In order to avoid the episodic character typical of survey courses, I link the material with a unifying theme of space travel and colonization. Although not every subject one would like to present can be related to the theme in a natural manner, many physical concepts lend themselves to such a treatment.[1] In fact, just getting students to understand how a rocket works (it does not push against the ground!) is a major achievement.

Students are generally aware of the possibilities of space colonization. They also know that an artificial colony, such as the long cylinder envisioned by O'Neill and others, would be set spinning in order to provide "gravity." Just how spinning something generates or mimics a gravitational force is not at all clear. Those who attempt an explanation usually do so in terms of "centrifugal force" which is believed to push outward against the wall of the structure. Asked to imagine that the wall suddenly vanishes, most students expect the inhabitants to fly off in a direction radially outward, away from the rotation axis. This is a misconception, and one surprisingly difficult to dispel (see Fig. 1).

14

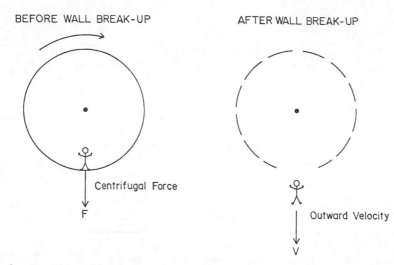

Fig. 1: Familiar but incorrect view of how rotation simulates gravity. The occupant of the rotating cylinder feels a centrifugal force F which pushes him against the wall. When the wall disintegrates, the centrifugal force ejects him into space in a direction radially away from the rotation axis.

There are many subtleties that are connected with rotation. An occupant of the space cylinder feels himself pressed against its internal wall and attributes this to a centrifugal force caused by its rotation. Let us now imagine everything in the Universe (planets, stars, galaxies, gas, etc.) removed one by one until only the cylinder is left. Is the cylinder rotating, and if so, with respect to what? The view of Newton, which will be adopted here, is that the rotation exists relative to space itself -- so called Absolute Space. The inhabitants would continue to experience centrifugal force.

An observer is postulated who is at rest in this space, and his frame of reference is called an "inertial frame." This refers to the fact that the inertial property of matter is manifest; a mass at rest remains at rest, or one in motion continues in motion in a straight line at constant speed as long as no force acts on it. A mathematical statement of motion in an inertial frame is Newton's Second Law

$$F = ma \qquad\qquad (1)$$

This is one of the simplest, yet richest, laws in physics. It states that a force F will cause a mass m to undergo an acceleration a, i.e., to change its velocity. If no force acts, F = 0, and then the acceleration a = 0. Zero acceleration means the velocity remains constant; the mass moves at constant speed in a straight line.

Let us consider what happens to the occupants of a space cylinder as viewed by an observer in the inertial frame, who also possesses the ability to see through the wall of the colony.

BEFORE WALL BREAK-UP   AFTER WALL BREAK-UP

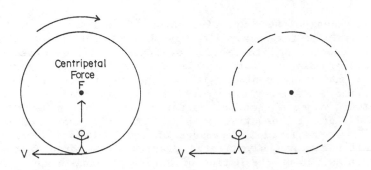

Fig. 2: The occupant of the rotating cylinder is acted upon by a centripetal force F exerted by the wall and directed radially inward. Just before the wall vanishes, he has a velocity V directed tangentially to the cylinder wall. When the wall disintegrates, the centripetal force F vanishes, so the victim continues to move with the velocity V in the tangential direction.

Earlier, we imagined the wall to vanish suddenly. Suppose this event is viewed from the inertial frame. What force acts on the hapless colonists after this disaster? None at all. Thus each colonist continues to move at the speed and in the direction he had at the instant the wall disintegrated. This direction is tangent to the wall not perpendicular to it (see Fig. 2). Students seem disinclined to accept this idea although it must be familiar to anyone who has taken a bicycle through a mud puddle.

Let us complicate the above thought experiment a bit in order to get at the nature of the force acting on an occupant of the space cylinder. Imagine that the wall continuously disappears and immediately reappears. When the wall vanishes, the inhabitant begins to move tangentially away, but instantaneously collides with the reformed wall, which accelerates him, and changes his velocity by deflecting it from a straight line. The continuous operation of this process results in motion in a circle about the rotation axis of the cylinder. The only force acting is the radially inward one due to interaction with the wall, i.e., it is a "centripetal" force. The fiction of a disintegrating and reforming wall may now be dropped, but the basic result remains -- the wall checks the inertial tendency of matter to move in a straight line and exerts a force which the occupants experience as something similar to gravity.

Before we can understand how the rotation simulates and differs from a gravitational field, we must investigate the effect of such a field on the motion of objects in it.

GRAVITY

Gravity has not been easy for anyone to understand and is still an active field of research. Although physics developed as a science in Ancient Greece, and the motion of falling bodies formed part of that study, gravity was not considered a force at all. Forces were exerted by movers in contact with things moved. Early scientific thinkers such as Aristotle rejected the notion that a force could exist between objects that were separated and not connected in any way. The term to describe such a situation was "action at a distance." It required the passage of nearly two millennia before Newton rather courageously treated gravity as a force (although he also was uneasy about the notion of action at a distance). Gravity was henceforth treated on an equal footing with any other force, e.g., a force exerted by pulling on a rope. Much as one can whirl a mass about on a string, the Earth could be pulled into a circular orbit similar to the one it occupies by a "superman" with a strong enough string. Interestingly, a steel cable stretching from the Sun to the Earth and having the diameter of the Earth is not strong enough, while the invisible gravitational attraction between the Sun and Earth is. Although we are conditioned to accept this as part of our world picture, one can see why it appeared preposterous.

With gravity treated as a force, it became subject to the laws of motion, in particular to Newton's Second Law. This led to a complete description of the motions of planets, satellites and comets that profoundly changed the notion of what man could comprehend. It also accounted for Galileo's earlier observation that all masses fall at the same rate. In terms of the laws of motion, this means that the force of gravity imparts the same acceleration to every object, independent of its mass. Traditionally, the acceleration of gravity is denoted g, and is numerically 9.8 m/s^2 or 32 ft/s^2. Because motion under constant acceleration is the least complex possible, very simple expressions can be derived for the velocity and distance moved. A mass dropped at time t=0 will have a velocity V and will have moved a distance S at time t given by

$$V = gt \qquad\qquad (2a)$$

$$S = gt^2/2 \qquad\qquad (2b)$$

Since the advertised topic of this lecture is baseball in space, let us use what has been discussed in that connection. A pitcher stands 60 ft from home plate and throws a fastball at a speed of 90 m.p.h. (For a traditional American sport, use of the metric system would be unpatriotic.) In order to model the pitch some simplifications will be made. First, air resistance is ignored (that means no curves or sliders). Second, the ball is assumed to be released in a direction parallel to the ground. The only force on the ball after it leaves the pitcher's hand is gravity, which pulls the ball down in the vertical direction. Since there is no force acting horizontally,

Fig. 3: The standard model of a pitch in a terrestrial baseball game. The pitcher releases the ball from a height of 6 ft at 90 m.p.h. and parallel to the ground. The ball crosses the plate 60 ft away at a height of 2.7 ft, within the strike zone (of normal athletes).

---

the acceleration in the horizontal direction is zero from the Second Law, so there is no change in the ball's horizontal velocity. Thus, the ball maintains its speed of 90 m.p.h. If the distance to the plate is divided by this speed, one finds the time of travel to be about 0.45 s.

Although there is no horizontal acceleration, there is certainly a vertical acceleration -- that of gravity, g. From equation (2b) we find that the ball will fall a distance of 3.3 ft after the time interval of 0.45 s. If it was originally released at a height of 6 ft above the ground, the ball will be at a height of 2.7 ft (6 - 3.3) when it crosses home plate -- nicely within the strike zone. The trajectory of the ball is shown in Fig. 3. This will be our "standard model" of a pitch.

THE ROTATING CYLINDER

One possible model of an artificial space colony consists of a cylinder with rounded endcaps and spinning around the longitudinal axis. This will be the model used here. Its diameter will be taken to be 1 km. This is not enormous as space colonies go in the speculations of more imaginative thinkers, but it is vastly bigger than anything humans have constructed in space or are likely to within this century. As discussed above, the inhabitants of the cylinder are constrained to move in a circle by the cylinder wall and are thus accelerated. The magnitude of the acceleration is

$$a = V^2/R \qquad\qquad (3)$$

If we want the acceleration at the cylinder wall to be equal to the terrestrial gravitational acceleration g, we need only use equation (3) to find the required value of V given the radius chosen. This is found to be 70 m/s. In order to see more clearly how fast the colony must rotate to produce the centripetal acceleration g, we will calculate how long it takes for one complete rotation to be made. In one rotation a point on the wall moves a distance equal to the circumference of the cylinder, which is, of course, $2\pi R$ = 3141.6 m. At the computed speed of 70 m/s the time required is nearly 45 s. This is the period of rotation, and it will be denoted T. With this period, the cylinder will make 80 rotations in one hour.

Before we contemplate a baseball game in the space colony, a somewhat simpler example will be useful. Although literal-minded authorities have debunked Galileo's Leaning Tower of Pisa experiment as being merely a fable, we imagine an exact duplicate of the Tower rising from the wall of the space cylinder. The Encyclopedia Britannica gives the height h of the Tower as 179 ft, which is approximately 55 m. We know that Galileo discovered that two bodies of different masses dropped at the same time strike the ground simultaneously. Using equation (2b) we can compute that the time it takes to fall the length of the Tower is 3.35 s. The position of a mass as it falls in equal time steps is shown in Fig. 4a. In the space colony the physics of the situation is quite different. The Pisa Tower is rigidly attached to the cylinder wall and rotates with it. If one marks a spot on the base and at the top of the Tower, both spots must return to the same points in space after one rotation period T (otherwise the edifice would be torn apart). Because the top of the Tower is a distance h closer to the center of the cylinder, the top lies at a radial distance of (R-h) from the center, while the base is a distance R away. Thus, the top moves in a circle of smaller radius than does the base, but in the same time interval. The speed of rotation of the top is $2\pi(R-h)/T$, which is 62.3 m/s compared to the 70 m/s found for the base at the cylinder wall. Note that if the Tower were exactly equal to the radius of the cylinder (500 m), then the speed of its top would be zero, since (R-h) = 0. (See Fig. 5).

In order to describe what happens when a mass is dropped from the Leaning Tower, we have taken the viewpoint of an observer in the inertial frame of reference who sees the cylinder rotate 80 times per hour.

When Galileo's space double releases the mass, there is no force acting on it. (The gravitational attraction of the mass to the rest of the colony mass is quite negligible.) No force means zero acceleration; the mass moves in a straight line and at the speed it had when released. This speed is the 62.3 m/s calculated above, and the direction is along the tangent to the inner circle described by the top of the Tower. The distance s that the mass moves before hitting the "ground", i.e., before striking the wall, can be found using only the Pythagoras Theorem (see Fig. 6):

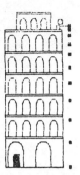 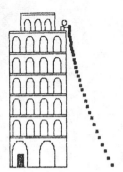

(a)   (b)

Fig. 4: The Leaning Tower of Pisa experiment of the Earth (a) and in
a rotating, cylindrical space colony 1 km in diameter (b). The most
significant difference is the horizontal motion that occurs in the
cylinder. The mass lands nearly 20 m away from the Tower, after
falling its 55 m height. Rotation, therefore, does not perfectly
simulate terrestrial gravity.

$$S^2 + (R-h)^2 = R^2$$

Since R and h are known, S is found to be 228 m. This is very
different from the terrestrial experiment, where the distance through
which the object falls is the Tower height of 55 m. The time it
takes to move a distance S = 228 m at the speed of 62.3 m/s is 3.7 s,
or about the same as the 3.35 s computed for the real Leaning Tower.
We have performed this calculation from the standpoint of the
inertial observer. A person standing at the base of the Tower will
see the motion in a very different way. During the 3.7 s it takes
the mass to move to the wall, the observer within the cylinder has
not remained stationary in space but has rotated with the cylinder
through an angle. In order to obtain his perception, we must take
into account his changing position in space. This is somewhat more
complicated than the calculations done so far and requires some grasp
of trigonometry. Here is where computer simulation can play a
valuable role. A program written for the Commodore 64 microcomputer
was used to simulate the motion of a ball from the Tower. The result
is shown in Fig. 4b. (A better programmer might have gotten the

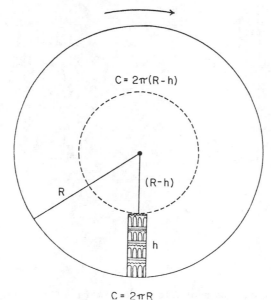

Fig. 5: In the time it takes for one complete rotation of the cylinder, the base of the Tower moves on the large circle of circumference $2\pi R$, while the top of the Tower moves on the small circle of circumference $2\pi(R-h)$. The height of the Tower is exaggerated in this picture to show the effect more clearly. Since the bottom moves a greater distance than the top in equal times, it has the larger speed.

Tower to lean. As it is, I spent more time than I'd like to admit getting the Tower graphics done.) It is clear that the object not only "falls down" as seen by the observer in the colony, but also moves to the side. This transverse motion is quite substantial, so that the object strikes the "ground" a distance of nearly 20 m away from the side of the Tower.

BASEBALL IN SPACE

The result of the Leaning Tower of Pisa experiment has important implications for common terrestrial sports played in a space colony. It should be emphasized that the inhabitant of the cylinder is not aware at all of its rotation. Such a person would attribute the curved path of the body dropped from the tower as due to a "gravita-

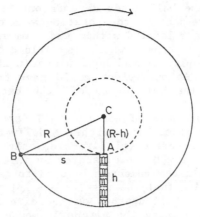

Fig. 6: The motion of an object released at point A from the top of the Tower as seen by the external observer in an inertial frame of reference. Just after its release, no force acts on the object, and it maintains the speed and direction it had at A. It thus moves from A to B through a distance S which can be computed from the Pythagorean Theorem for the right triangle ABC. At B an observer within the cylinder sees the object strike the "ground," which is, of course, the internal cylinder wall.

---

tional field" rather more complicated than the one familiar to us on Earth. No doubt, after long experience athletes would be able to function in this environment, while visiting teams from Earth would find themselves at a disadvantage.

    To see what one has to deal with consider a high pop-up that is hit directly upward at the plate. If this occurs in a game played on the Earth, the catcher has about 5 s to field the ball. The motion of the ball is symmetrical; it takes 2.5 s to reach maximum height and 2.5 s to return. We use equation (2b) to find the distance traversed in 2.5 s and obtain 100 ft. Equation (2a) gives the speed of the ball when fielded as 80 ft/s (about 55 m.p.h.). This is also the speed of the ball when it left the bat.

    On a playing field in a space colony the outcome of a pop-up is drastically different. As before, we let the ball be hit directly upwards (in a cylinder this is along a radius) at an initial speed of 80 ft/s. Since we have seen that "gravity" within the space colony has both radial and transverse components, we anticipate that the ball will drift horizontally away from the plate. The computer simulation verifies this behavior. The ball rises to approximately 100 ft in 2.5 s, just as on Earth. But when the ball is fielded, it is certainly not by the catcher; it has moved about 90 ft from home plate!

Exactly where the ball lands depends on the orientation of the playing field with respect to the rotation axis of the cylinder. If two baseball fields are located within the same space colony, but are aligned differently, a pop-up which is fielded by the shortstop in one could be out of play behind home plate in the other. And this was just a pop-up. An outfielder would have to take into account more factors since the ball comes off the bat at some arbitrary angle.

We end this discussion of how baseball fares in a space colony by reconsidering our model of a pitcher's delivery. Recall that the ball is released in a horizontal direction only (in a cylinder this is along the tangent) at 90 m.p.h. and at an initial height above the ground of 6 ft. Again the computer is used to calculate the path of the ball. You may find that the results are not intuitively obvious. The trajectory of the ball depends on the direction in which it is thrown. If the line joining the pitcher's mound and home plate lies in a direction parallel to the rotation axis of the cylinder, the ball behaves much like one thrown on the Earth. It crosses the plate within the strike zone, and takes the same time to traverse the distance. The difference is that the ball deviates from a straight line by 1 3/4 inches when it crosses the plate. Thus, a hard fastball has turned into a curve. (Don't confuse this with a curve thrown on Earth, which depends upon the flow of air around a spinning ball.)

If the playing field is rotated by 90 degrees, so that the pitched ball moves in the direction that the cylinder turns, the behavior is quite different. Although the delivery is identical, the ball never makes it the 60 ft to the plate and strikes the ground short by nearly 10 ft. If the pitched ball moves in the direction opposite to that in which the cylinder turns, the ball hardly drops at all and crosses the plate at a height of 5 1/2 ft but without curving.

All these results can be computed by the same technique. In each case the external observer, who is assumed to be in an inertial frame of reference, sees the ball move in a straight line at constant speed as soon as it leaves either a player's hand or the bat. This straight line path is then viewed from the point of view of a player on the field in the rotating cylindrical space colony. A bit of geometry and trigonometry suffice to compute the trajectory of the ball. The non-intuitive nature of some of the results suggests that traditional terrestrial sports may have to be considerably modified, or that athletes will have to develop new instincts.

COMMENTS

Some people are puzzled by the example of the high pop-up. Why, since the ball was hit in the radial direction, does it "fall" to the ground and not cross the cylinder entirely. The answer may be

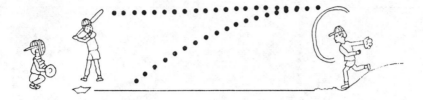

Fig. 7: The same delivery that produced a strike in a terrestrial baseball game (see Fig. 3) follows different trajectories in the space cylinder depending upon the orientation of the playing field. If the ball is thrown in the same direction as the cylinder turns, it never makes it to the plate. If it is thrown in the opposite direction, it sails over the batter's head.

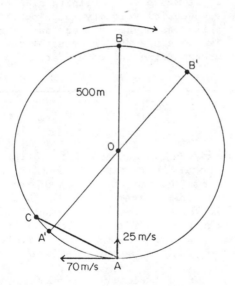

Fig. 8: The player at A wishes to throw the ball to the player opposite him at B, and propels the ball along the line AB. Unfortunately, the ball already has a large tangential velocity, which causes the actual path to be the straight line AC (as viewed by the observer in the external inertial frame). During the time it takes the ball to move from A to C, player A has changed position, and is at A'. Thus, rather than moving the 1-km distance intended, the ball lands only about 25 m from player A's feet.

best appreciated by changing the question slightly. Is it possible

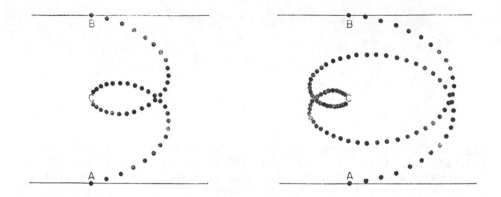

Fig. 9: In principle it is possible to throw a ball across the 1-km diameter of the cylinder from A to B by giving it a sideways velocity equal and opposite to that of the cylinder wall. Since this means a speed of over 150 m.p.h., it may prove something of a chore. In addition to the initial tangential velocity, a velocity in the radial direction AB must be supplied such that B makes one or more complete revolutions in space. Two possible trajectories of the ball as seen by player A are shown corresponding to two different radial velocities.

---

for two people at diametrically opposite points to throw a ball back and forth? The situation is illustrated in Fig. 8. As usual, we first view the activity from the inertial frame of reference. When A has the ball, it is moving instantaneously to the left at a speed of 70 m/s (over 150 m.p.h.), which is the rotation speed of the cylinder wall. If A simply propels the ball in the direction AB, the initial tangential velocity remains unmodified. For definiteness, let A throw the ball along AB with a speed of 25 m/s, which is approximately 55 m.p.h.

The combination of the two velocity components causes the ball actually to traverse the straight line AC. During the time the ball is in flight, the players change their positions in space to A' and B'. Thus, if one mentally transforms to the perspective of player A, it is clear that his intention of throwing the ball to B is frustrated by the rotation, and it strikes the "ground" only about 26 m from where he stands. This situation is precisely that of the pop-up discussed earlier.

How then does one play catch with a friend on the other side of the cylinder? First, it is necessary to cancel the original tangential velocity of the ball. This is done by throwing the ball at the same speed the cylinder is rotating but in the opposite direction and at the same time giving it a component of velocity

along line AB.   The inertial-frame observer then does see the ball
move along line AB.   The speed along AB must be such that B returns
to his original position when the ball gets there.   Thus, the ball
has to move the length of the diameter (1 km) in one rotation period
T (about 45 s).   This means, it should be thrown at 22 m/s (about 50
m.p.h.).   If the ball is thrown at exactly half this speed, B will
have made two complete rotations and be in position to make the
catch.   Player A can therefore throw the ball at a series of
velocities corresponding to the times for B to make any number of
complete rotations.   The trajectories of the ball as seen by player A
become increasingly complicated as it appears to make multiple loops
around the central axis before finally being caught by B.   Such a
case is shown in Fig. 9.   Playing catch in a cylinder in this way is
no simple matter!

REFERENCES AND FOOTNOTES

1.   See for example, C. H. Holbrow, "Physics of Living in Space," Am.
J. Phys. 49, 725-32, (1981); or Jay S. Huebner "Teaching about
the Colonization of Space," Am. J. Phys. 47, 228-31, (1979).

FROM BUS STOPS TO SPACE COLONIES:
Reflections on Using Space—Colony Concerns
in Teaching Architectural Design

Patrick D. Hill, AIA
Professor of Architecture
California Polytechnic State University
San Luis Obispo, CA 93401

ABSTRACT

In the course of a formal academic education, the student of architecture learns to synthesize many, often conflicting issues into a
satisfying and holistic solution. Space colonization presents the
student with a new and rich set of potentials and constraints to deal
with. Its scale and complexity as a problem, however, shows them
their limits as designers. Trained to be a leader, coordinator and
guiding visionary of building form, what is the architect's place in
the eventual colonization of outer space and how should the architect
be prepared to contribute?

INTRODUCTION

For the beginning student of architecture the design of a bus
stop is a challenging and useful exercise. The convincing design of
something as grand as a space colony with its variety of disciplinary
concerns and its intricacies of technology is probably beyond the
capability of any individual. Nevertheless, the ideas of space
colonization have generated a number of smaller design projects that
have motivated my architecture students and resulted in interesting
proposals. Moreover, those who have considered the scope and demands
of an overall colony design become keenly aware of the limits of what
a single architect can hope to contribute to any large-scale, technologically complex project.

THE PROBLEM OF DESIGN

Design has always involved making decisions. Designs of necessity have usually been fairly direct solutions to specific needs involving few alternative choices and mostly dictated by the function they serve. Or so it would seem. Look at a doorknob. Its height, location, projection from the door, and size all have to do with the human hand and figure, to be sure, but the choice of material used, its color, texture, finish and shape are really up to the designer. It can be a lever just as well as a knob; the knob might be square instead of round, or oval. It might be a shiny brass finish or a dull black. It can be placed at the edge of the door or in the center of the door. The essential function or arrangement of the knob, to release a latch to unlock the door, however, remains the same.

Normally, as the object or thing that is being designed becomes more complex, with more parts, the number of choices and decisions to be made increases. The car for instance, although designed in many different styles and shapes, embodies one essential arrangement: a shell to enclose the occupants, a propulsion system and wheels. Yet the number of decisions, economical, technical, and aesthetic needed to produce a particular design is enormous.

Limitations arise only in part from the physical nature of the world; more stem from considerations of economy and style, purely matters of choice. Quoting David Pye in The Nature of Design "All the works of man look as they do from choice, and not from necessity."[1]

The more complex a design problem becomes, the more decisions need to be made about many more possible choices. The designer must find a way to organize decision making, or chaos ensues and the most appropriate design, or desired arrangement of parts to achieve a particular result, is not achieved. In making design decisions the more analytical designer might use a systematic method to arrive at an acceptable choice. This process would involve weighing each choice against specific criteria of performance: Cost effectiveness, ease of building, or aesthetic desirability. By repeatedly examining possible choices or variations against common criteria, the designer hopes to arrive at the most acceptable alternative. The more intuitive designer hopes to imagine a solution that resolves all functional and economic concerns within a desired image or form, the image being a visual, compositional order that aids in making decisions about color, form, texture and material. Usually the good design lies somewhere between the intuitive response and the analytical search.

In designing architecture, the architect must understand the needs of the building, the people using it and the environment surrounding it. These factors guide the particular arrangement of the structure's walls, spaces, windows, stairs, to name but a few of the parts. A concert hall must suit the needs of the people sitting in it but must also work well both as part of an acoustical system and as an enclosing system. Its exterior shape must also suit particular climatic conditions as well as fit into an urban or rural visual setting. Buildings that fit into the countryside do not normally look right in a city and vice versa. Probably everyone would agree that a house should look like a home and not a gas station. Choices, decisions and compromises abound.

THE BUS STOP AS AN EXERCISE IN DESIGN

To train architecture students to cope with this dilemma of design, it has been traditional to assign them the task of developing a proposal for a "bus stop." The task has the advantage that it embodies many of the problems encountered in building design in a small exercise that can be dealt with in a short time. It allows exposure to a complete process: the need to gather facts about wind, sun, rain, site conditions, transit authority requirements, human physical requirements (ergonomics); the need to address a range of design issues: human comfort, security, protection from the elements, structure, site and context, appropriate image, public art, identification with function, the need to imagine alternative solutions and to select and refine the best or most desirable choice. In short, it is an exercise that begins to develop the student's understanding of dealing with real problems in a systematic conceptual way. The aesthetic issues of visual concept, visual vocabulary (proportion, balance, harmony, rhythm) and scale must all be considered to achieve a pleasing yet functional human shelter. Subsequent exercises and years of design experiences train the architect to deal with ever more complex problems and the resolution of many more multiple concerns in a holistic way. Eventually the student develops the intuitive attitude and skill to realize that good architecture is an integrative discipline, requiring the designer to solve many problems with a single design.

THE SPACE COLONY AS AN EXERCISE IN DESIGN

How does one even consider using the scenario of colonizing outer space as a studio-learning study? This is a design exercise that goes far beyond the typical range of building concerns. It deals with much broader social, economic, technical, psychological and physical problems. The issues involved are quite unlike the earthly problems encountered in building design. A single designer working alone would have to have training and experiences in many fields to be able to make sound design decisions that would interact successfully with many other aspects of the eventual solution. It certainly is a multi-disciplinary design problem.

Many of the elements of the typical architectural design process were evident in the 1975 NASA/Ames summer space project to design an orbiting city in space for 10,000 people.[2]   There were multiple issues: specific, clearly stated criteria or needs, various alternative solutions and concepts proposed, and the selection and final, if hasty, development of one of the design alternatives. But the use of the space colony as a case study to train the novice student architect to deal with building design is very debatable.   It certainly is more appropriate as a topic in architectural design for graduate students than for undergraduates.

A Graphic Visual

Nevertheless, when I came back to teaching from the 1975 Ames Study, the excitement of the idea of space colonization and how that excitement might be used to teach design were on my mind and eventually fostered several projects. An early exercise, even before the results of the Summer Study were published, was to have a class design as a short sketch problem a front cover for the final study publication.[3]   The strongest solution, from my perspective, was sent on to Ames as a suggestion. While not used directly, it was the inspiration for the frontispiece of Chapter 8 in the report. This was a useful exercise in motivating the class to design using principles of composition, graphic presentation and conceptualization. Overall results were of a very high quality.

A Comprehensive Design

In late 1977 a group of five senior design majors in our four-year B.S. program in Architecture came to me interested in doing a space-colony design study as a senior project. I guess my reputation was getting around. I gave them a lot of the data I had gathered from the study, and they set about reviewing it and came back to me, as their project adviser, with a proposal. I was surprised that they had decided to take the toroidal structure proposed by the 1975 Summer Study group and named "The Stanford Torus" as a given and to develop a proposal for a habitable community within it. I guess they too realized how large would be the problem of trying to design a complete colony system from scratch. The result of their study, titled The Humanization of an Extra-Terrestrial System[4] is not strong visually or architecturally, but is a successful examination of alternative arrangements of housing and physical density. They demonstrated that the alloted volume for the interior of the Stanford Torus was enough to make a strong urban-like setting for living and working, something I had wondered about when the final dimensions were selected at Ames.

Designs of Parts of a Space-Colony System

Several other projects at both the graduate and undergraduate level have been done since that time. Most, in some cases at my insistance, moved away from doing an overall systems design of a space colony and focused on a more realizable architectural problem. Several dealt with a lunar-mining facility.

In 1980 a graduate student, again after investigating the idea of a space-colony design as a master's thesis project, decided to design a large 100-person orbiting research lab. The final thesis had extensive analysis, but little to offer as a serious design. The student concentrated too much time on gathering and analyzing data, which was important, but then had to be hasty in synthesizing a proposed solution. It needed more time for development and refinement. The process was incomplete. The selected concept needed more testing against technical requirements (much more necessary for a space station than for a building) and further refinement in order to be valid.

About the same time, a more energetic and visually mature designer in his last year of our five-year BArch program decided to take on a lunar research base as a year-long fifth-year project. He went through a fairly complete technical review of the problems involved and built a good study model at a large scale to test structure, geometry and form. The final presentation model and documentation were of a high quality. It was, however, not a space colony. The project had limited formal architectural goals and was well developed as such.

In 1982 I gave another short sketch problem to stimulate design decision-making in a fourth-year undergraduate studio. This time I had the class take the Stanford Torus as a given, develop a concept for the interior, and then focus specifically on designing the entry and exit area from a "spoke" into the community level.[5] They were to design a space or sequence of spaces to accommodate large numbers of people leaving to the exterior in shifts of up to 2,200 people at a time, to handle freight coming to the interior from the outside, as well as to accommodate new arrivals or visitors from Earth. Lounging and waiting, transportation, security and rescue concerns had to be addressed. It was a very Earth-like design problem in the context of a space colony, and the students thoroughly enjoyed the project. They did library research on similar Earth-like situations such as airports and large hotels. The interior studies were intriguing and filled with energy. This turned out to be an exciting exercise. In fact, several students asked to continue the problem further, into the housing area, for the remainder of the quarter.

And lastly, as recently as Fall 1984 another graduate student has undertaken, for the first time, a complete study of a space-colony system for his master's thesis project. It has been a learning situation for the student from the start. He has made a gallant effort at tackling some major technical consequences arising from the requirements of human physiology: the need for pseudo-gravity and for radiation protection. The proposed system uses

Tsiolkovsky's idea of tethered masses to achieve a very slight pseudo-gravity and is placed within the Earth's Van Allen belt for radiation shielding, a solution about which technical questions might be raised. The inhabitants are to be individually compartmentalized, relying on self-selected atmospheric and holographic created experiences for environmental stimulation. It poses some exciting and new possibilities. Many concerns regarding human psychological response, social interaction and governance, isolation and confinement (especially the paranoia people reportedly experience in a totally artificial, man-made environment), economic feasibility, and physiological response to very low gravity have yet to be addressed. Again, we see the problems of the single designer grappling with a complex, large-scale problem when his or her experience and skills lie in a limited number of traditional areas.

SUMMARY AND CONCLUSION

The problem of designing a complete system for permanently housing large numbers of people in space goes beyond the typical capabilities and experiences of even the most skillful architectural designer. And why should the traditionally Earth trained architect necessarily be the one to propose design solutions for space colonization? It took no architect to do Skylab. To be really instrumental in an overall system concept for colonizing space, the architect would have to become, as in many large Earth projects, more of a coordinator and facilitator of several other design professionals, or a collaborator in a design team where the architect shares ideas with others in developing an optimum solution. Perhaps a good systems engineer with some personnel management training could do the job just as well. Perhaps an engineer's "design" ego would be less likely to get in the way of making the most efficient design decisions.

Economics, space transport and orbital mechanics, space engineering and physics, sociology, physiology and psychology play the major roles in designing an overall system for colonizing space, not the architectural student's preoccupation with form, image and social art. The ivory tower designer of The Fountainhead[6] will play no role in our going into space. There is current evidence of this in NASA's recent hiring of two architects at the Ames man/machine laboratory at Moffett Field to work on interior planning for the US/Europe orbiting manned station. NASA engineers are coordinating the efforts of the architects, a reversal of their traditional roles. Architects, it seems to me, will have to give up being prima donnas, and become members of a much larger team and effort than they are used to. Architectural education that teaches designers a holistic, integrative approach to design is still valid, but it must broaden its perspective to include the input of more technical and physical constraints than it is used to. Only if a discipline evolved that could truly educate architects in all of the physical and social problems of designing and making decisions on so many interrelated systems issues, would space-civilizing architects be capable of taking their traditional place as design leaders and coordinators.

REFERENCES AND FOOTNOTES

1. David Pye, The Nature of Design, (Reinhold, New York, 1964), p. 11.
2. Space Settlements: A Design Study, NASA SP-413, edited by Richard D. Johnson and Charles Holbrow, (Government Printing Office, Washington, DC, 1977).
3. Space Settlements, op. cit.
4. G. Smith, et al., "The Humanization of an Extra-Terrestrial System," Senior Project, California Polytechnic State University, San Luis Obispo, 1977.
5. The Stanford Torus is entered through its hub which is connected to the living areas in the ring by 865-meter long, 15-meter diameter "spokes."
6. Ayn Rand, The Fountainhead, (Bobbs-Merril, New York, 1943)

SPACE IN THE CLASSROOM

William E. MacDaniel
Space Settlements Studies Project
Department of Sociology, Niagara University,
Niagara University, NY 14109

ABSTRACT

As we enter into the space age we must realize that our space activities are likely to constitute germinal input to an extraterrestrial society and its culture which will be uniquely different from any found on Earth. It is vital that the current generation of students have the opportunity to learn as much as possible about the nature of the changes which the space age will necessitate in both terrestrial and extraterrestrial society and culture, and the impact that such changes are likely to have upon career and lifestyle. To these ends I introduced a space related course into the Niagara University curriculum with the two goals of fostering student understanding of the sociocultural forces which shape their lives and of helping to prepare them for life in the space age. This paper describes the course, its difficulties and its prospects.

INTRODUCTION

Sociology 290, Living in Extraterrestrial Space, was offered for the first time in the fall 1984 semester. The course was introduced to foster student understanding of social and cultural forces and to help students prepare for life in the space age.

Difficulties arose early. Low enrollment nearly resulted in the course offering being cancelled. Those students who did enroll in the course were of mixed background in academic major, class level, exposure to knowledge about space and the sociological concepts which were to be employed, and in level of interest in the subject matter. It became apparent during the progress of the course that emphasis had shifted from the sociocultural aspects of space living to the technological and physical requirements for space survival. Other difficulties involved uncertain delivery of requested audio-visual aids and student reluctance to engage in outside reading of space-related materials.

In spite of these difficulties the students who participated in the initial offering of the course indicated that they found the subject matter interesting and the course to be both informative and entertaining. The overall goals of the course were achieved, and it will continue to be offered in the future although with extensive revision.

RATIONALE FOR THE COURSE

The natural state of man is in a constant state of war, where every man is enemy to every man. In such a state there is no place for industry, no navigation, no commerce, no architecture, no mechanics, no knowledge of the Earth, no arts, no account of time, no letters, no society. There is continual fear and danger of violent death -- the life of man is solitary, poor, nasty, brutish and short. (Thomas Hobbs on the natural state of man, Leviathan)

Today we are on the verge of expanding the reach of humanity to the stars. We have walked on the Moon; we are embarked upon a program to establish a permanent human presence in cislunar space, and we possess the technological knowledge requisite to establish permanent human communities on the lunar surface and in orbiting habitats. Within the next few decades we are likely to have thousands of humans working in space on a routine basis, and members of the younger generation of today are likely to be participants in the genesis and early evolutionary struggles of an extraterrestrial civilization.

What will be the natural state of extraterrestrial man? Will people live in a Hobbsean world of strife and misery, in a world which is akin to that which is found on Earth, or in a utopian society of their own deliberate creation? Whatever the state of extraterrestrial man, it will be the product of their efforts to survive in the uniquely hostile extraterrestrial environment through activities that are organized in accordance with social norms and patterns that become institutionalized within the evolving culture and social structure of space. It is necessary, therefore, that those who will participate in the humanization of space have knowledge of, and appreciation for, the social and cultural mechanisms through which humans regulate and organize their behavior to facilitate survival in a hostile environment.

Our present generation of students will be part of the movement toward humanization of extraterrestrial space. Like it or not, space will increasingly become an important resource for exploitation and utilization and will become an active arena for economic, political and military competition, and for conflict between the nations of Earth. Many members of our present student generation will be active participants in future space activities through their occupational and career pursuits; the remainder will be passive participants through the actions of their governmental and economic leadership. No one in the industrialized world will be able to avoid the impact of the space age, because it will bring about profound changes in world view, lifestyle, culture and society on Earth, and development of a unique human society and culture in extraterrestrial space. It is necessary, therefore, that students in all academic disciplines be afforded the opportunity to acquaint themselves with the nature of extraterrestrial space, the opportunities which the space age holds forth to them, and the impact that our reach for the stars is likely to have upon their lives.

Such was the general tone of the rationale that was employed to gain curriculum committee approval for a new sociology course, "Living in Extraterrestrial Space," to be offered initially to Niagara University students in the fall semester of 1984. New course offerings, however, often have a way of going awry.

GOALS OF THE COURSE

"The best laid schemes of mice and men. . ."

As originally conceived, "Living in Extraterrestrial Space" was to be an elective course open to all students; its goals were both sociological and practical:

- To enable students to understand the manner in which social and cultural forces circumscribe human behavior, the manner in which these forces are related to the environment in which human groups must struggle for survival, and the manner in which these forces can contribute to, or detract from, the human condition.

- To help prepare students for life in a rapidly changing world through acquisition of knowledge concerning a significant social phenomenon which holds the potential for greatly influencing the human condition -- the industrialization, utilization and humanization of extraterrestrial space.

Essentially, the concept of living in space was to be used as a medium through which to emphasize the impact of social and cultural forces upon human behavior and well-being. Weightlessness during orbital flight, a particularly unique aspect of the extraterrestrial environment, was to be the key element for bringing the attention of students to the relationship between environment and social and cultural norms. The extent to which gravity is unobtrusively imbedded in Earth norms and behavior patterns would be emphasized, and students would be assisted in developing ideas of how norms would differ in an environment of which gravity was not a part.

Consider some examples. In an orbital habitat the feet lose their primary earthly function -- providing mobility -- because of weightlessness; dirt and filth do not settle to a floor to accumulate on and around the feet; people do not necessarily share a common vertical alignment in weightlessness, and they may find face-to-feet juxtaposition to be as common as face-to-face when engaging in social interaction. What are some alternatives to mobility for which feet might be used in space? How might norms and attitudes toward feet and bodily juxtaposition differ in space?

In order to take this approach it was first necessary to provide students with a working knowledge of the extraterrestrial environment and its impact upon the human condition. Accordingly, the first portion of the semester was to be devoted to discussion of the extraterrestrial environment. Beginning with a brief historical survey of our efforts to study and utilize space, we were then to move into the gravitational system (see the Appendix for a complete topical outline of the course). This was to be followed by discussions of space industrialization and utilization, the political implications of space utilization, and the physiological effects of weightlessness upon the human organism. This material was to be covered during the first few weeks of the course.

Having grasped the essential elements of the extraterrestrial environment that were likely to have major impacts upon sociocultural development in space, students would devote the remainder (majority) of the course to assessing terrestrial and extraterrestrial sociocultural implications of space humanization. Terrestrial implications were to include new opportunities in terms of jobs, resources, new products, the manner in which these might affect sociocultural patterns on Earth, and problems which might arise from space utilization. Extraterrestrial implications involved study of problems for which sociocultural, rather than technological, solutions would be needed. Other topics were to include extraterrestrial lifestyles, relationships with Earth, government, dependence upon Earth for life-support, deviance, social control, social class structure in space, unique functional and dysfunctional cultural traits that might develop in space, and identification of Earth cultural traits which might be dysfunctional if carried into space by immigrants. Finally, discussion was to elaborate and distinguish between the concepts of space colony and space community.

Two papers were to be required. The first would relate each student's academic major to activities in space. The assignment was to provide an analysis of the occupational opportunities toward which study in an area of major academic concentration (the student's own) provided access, and how these might be influenced by developments in extraterrestrial space. The second paper was to be an end-state driven process scenario that would trace events and processes which might lead from the present state of space activities to some space-related future.

Required reading consisted of two books: The High Frontier by Gerard O'Neill[1] and Space Settlements: A Design Study edited by Richard Johnson and Charles Holbrow.[2] Students were also expected to read selections from other sources as well as from the instructor's personal materials that had been placed on library reserve.

Extensive use was to be made of video-tapes, slides and films from NASA.

ACTUAL OUTCOMES OF THE COURSE

. . . go aft astray and leave us to naught but grief and
pain for promised joy . . ."   (Robert Burns)

The initial problems were administrative. Curriculum committee approval was not forthcoming in time to enter the course description in the University Catalogue. It was necessary, therefore, to provide faculty colleagues with the course description and to ask that they call it to the attention of their students prior to registration; notices were also posted around the campus. Next, in the class schedule for registration the title of the course was abbreviated to "Exteri Living." As a result for fall-semester registration this offering became a "mystery course" about which very little was known. Naturally, the enrollment was low, six students, and this engendered conflict with the dreaded "rule of ten."

According to the "rule of ten," courses that do not have at least ten students will not normally be offered. It was possible, however, to convince the Dean that the unusual circumstances of the course schedule listing and catalogue problem warranted waiver of the rule. The course was allowed to be offered for the six students who, within the first week of the semester, managed to recruit four more enrollees.

The result was a mixed bag of students. Areas of major academic concentration included biology, nursing, foreign languages, theater, political science, English, business administration, and travel-transportation-tourism. Class level ranged from sophomore through senior, and interest in the course topic ranged from "gung-ho" enthusiasm (one student) to "what's going on here?" and "where am I?" (several students). When queried as to why they enrolled in the course, one student (a former editor of the campus paper for which I had written several short items on space topics) indicated that he wanted to pursue a space-related career and that he was very interested in learning all that we had to offer about space. Another student (who had been in one of my classes the previous semester) remembered an announcement about the space course, thought that it sounded interesting, and liked the way the instructor performed in class. Most of the others indicated that they needed a social science elective; had seen a notice about the course and thought it "might" be interesting; the time schedule met their needs; they had heard that the instructor was a "not bad" kind of teacher; and/or they had a friend in the class. Intense interest in learning about space was obviously not a strong motivating factor.

Two problems became immediately apparent: for the most part the class was almost totally unaware of what had been happening in space and of the potential for space utilization; and most members of the class had not been exposed to how sociologists think about the human condition or to the sociological concepts around which the sociocultural aspects of the course were to be centered. The course had to be restructured. The space education of the class had to begin from scratch and required considerably more time and attention than I had planned to give to the non-sociocultural aspects of living in space. It was also necessary to devote more time to explaining basic concepts of sociology.

The written assignments also engendered some difficulties. The purpose of the first paper was to encourage students to associate space with the future careers for which they were being prepared by study in a particular academic major. It so happened that some members of the class had very little idea about the constraints that study in a particular field place upon career opportunities, and some did not know what employment opportunities were associated with their particular major. In addition, it was necessary to know something about space in order to write the paper, and the time slippage prevented acquiring such information before the paper was due. It was necessary to delay the due-date for the paper. This pushed the first written assignment into the time period set aside for research of the materials required for the second assignment (the end-state driven scenario). In the end this made no difference because most of the class read little from the materials that had been placed on library reserve or from any other source. Several students indicated that they customarily operated on a "crash basis" and did not prepare any written assignments until the shortest possible period of time immediately preceding the deadline. Finally, no one in the class except the theater major knew what a scenario was or how to go about writing one. It was necessary to devote almost an entire week to discussion of writing an end-state driven process scenario. In spite of this, some of the papers that were written were not scenarios; they were library research papers of the type that are common in many courses and which do not usually require imaginative extrapolations into the future.

Everyone in the class reported favorably on The High Frontier, but Space Settlements was regarded as being too technical by some of the students. The latter is not a highly technical treatise on space settlement design, but it does require that its readers have minimal familiarity with certain concepts of mathematics and physics. Considering the varied background of the students in the class, it is not surprising that some had a very little mathematical exposure and had never taken a course in physics.

The use of audio-visual aids also led to some difficulties. The primary purposes in using such aids were to stimulate student interest in space, to provide information concerning past and current space activities, and to convey a "feeling" for what it is like to live in weightlessness. Our campus telecommunications center had a few videotapes related to space, but most of the materials I planned to use were films to be obtained from NASA. Some of the films were quite old and in rather poor condition; some appeared to be developed for high school classes; there was a considerable amount of redundancy in the film content, and substitute films were occasionally sent instead of those requested. In addition, NASA apparently has not produced many recent films (the most current catalogue was dated 1982) and much of the available material was quite dated.

CHANGES FOR THE FUTURE

In retrospect, it appears that much of the difficulty of this course resulted from the instructor being caught in the "overadvocacy trap," which is probably not an unusual problem for space enthusiasts. I expected that students who enrolled in the course would share my interest and enthusiasm for learning about space and that they would be willing to devote some of their time and effort to pursuit of this interest. Most of the students, on the other hand, wanted credit for a social science elective in their academic programs, were not very interested in pursuing material that was not clearly and directly related to their academic majors, and thought that this course would not be overly boring. Instructor and students were operating at cross-purposes on the basis of differing assumptions about the course.

In the revised course I will shift emphasis from extraterrestrial society to the impact that the space age will have upon terrestrial society. Space utilization, industrialization and commercialization will become dominant themes with a sociological perspective stressing how these will influence terrestrial lifestyle, world view, military and political arrangements, new and changing career opportunities, and the economy. Attention will also be given to educational requirements for living in the space age. The written assignment involving the end-state driven process scenario will be eliminated, and the career/occupational paper assignment will be expanded to include a section relating to future developments in a specific career area as a result of expanding capability for use of the extraterrestrial environment. A major section relating to the sociocultural impact of the extraterrestrial environment upon human social behavior will still be retained. The course title will be changed to "Space Age Living" which I hope will be more meaningful to students and more easily abbreviated by the registrar. I believe such revisions will orient the course more closely toward the needs and expectations of undergraduate students while still retaining the disciplinary integrity of an elective sociology course.

The NASA films were duplicated on videotape and will be edited in order to reduce redundancy and to produce a few thirty-minute tapes directly related to topics under discussion in the course.

The lack of knowledge about relationships between careers and academic majors was somewhat surprising but may not be unusual among undergraduate students who have matriculated in the liberal arts and social sciences. The course revision will include a section dealing specifically with the sociocultural relationships between career and education, and with the manner in which these relationships are likely to change in the space age. In addition, the research requirements for the written assignment will force independent student investigation of these factors.

Gerard O'Neill's book, The High Frontier, will be retained to introduce students to ideas about the potential that space holds for human utilization. However, the varied background of the students militates against retaining Space Settlements. A search is currently under way for a text which will pique the imagination, be compatible with the interest and knowledge background of a diverse group of undergraduate students, and which will serve as a reference source for ideas and further pursuit of knowledge. Much material of this nature is available in disparate sources, but students in the course have exhibited a reluctance to pursue these voluntarily. In the event that a suitable text cannot be found, a required reading list of materials on library reserve will be substituted for the second text.

Finally, the course goals will be modified to correspond more adequately to the needs of the diverse backgrounds of students who are likely to elect to take this course:

1. To facilitate understanding of the manner in which sociocultural forces affect and constrain life-chances.

2. To provide understanding and knowledge which will assist in achieving satisfying lifestyles in the space age.

In spite of a variety of difficulties, this first offering of "Living in Extraterrestrial Society" is regarded as a successful effort to introduce space studies into the Niagara University curriculum. I found the course to be both challenging and exciting. Students indicated that they had begun to think about the sociocultural aspect of the world in different ways, and that they had acquired a great deal of information about space and its potential for the future. I feel that the goals of the course were achieved, although not with the depth that I had anticipated. The revised course will be retained as part of the sociology curriculum.

REFERENCES AND FOOTNOTES

1. Gerard K. O'Neill, The High Frontier: Human Colonies in Space, (Doubleday & Co., Garden City, NJ, 1978).
2. Space Settlements: A Design Study, NASA-SP 413, edited by Richard D. Johnson and Charles Holbrow, (Government Printing Office, Washington, D.C., 1977).

APPENDIX
SOCIOLOGY 290: LIVING IN EXTRATERRESTRIAL SPACE
Course Prospectus (Fall, 1984)

Course Goals

I.   To understand the manner in which social and cultural forces
regulate and circumscribe human behavior, the manner in which these
forces are related to, and arise from the environment in which human
groups must struggle for survival, and the manner in which these
forces can contribute to, or detract from, the human condition in
that environment.

II.   To help prepare for life in a rapidly changing world through acquisition of knowledge concerning a significant social phenomenon
which holds the potential for greatly influencing the human condition
within the next few generations: the utilization, industrialization
and humanization of extraterrestrial space.

III.   When you have completed this course, you will have a greater
understanding of the sociocultural forces which shape and control
human lives, and you will have the potential for using this understanding in assessing alternative courses of action related to
personal behavior and to human expansion into the realm of extraterrestrial space.

Written Assignments

I.   Write a research paper in which you examine the implications of
future advances in space utilization, industrialization and humanization for the employment opportunities and working conditions in those
career areas for which your major area of academic concentration constitutes preparation.

II.   Write an end-state driven process scenario in which you identify
some future, space-related condition (the end-state) and trace the
process (sequence of events) by which we progress from the current
state of space activities and capabilities to that end-state.

Required Reading

I.   The High Frontier: Human Colonies in Space, Gerard K.O'Neill,
(Garden City, NJ: Doubleday & Co. 1978).

The purpose of this reading is to provide students with a sensitivity for the human potential in space and to pique their imagination. Reading is to be completed within the first two weeks of the semester.

II. Space Settlements: A Design Study - NASA SP-413, Eds. Richard D. Johnson and Charles Holbrow, Washington: National Aeronautics and Space Administration, 1977).

The purpose of this reading is to acquaint students with some of the more technological aspects of space settlements and to provide a reference source.

Course Outline

I. From Sputnik to the Space Shuttle:
  a. A brief historical survey of space flight advances.

II. The Extraterrestrial Environment:

  a. The Earth/Moon gravitational system.
  b. Mechanics of bodies in orbit.
  c. Weightlessness.
  d. Lunar and orbital habitat designs.

III. Space Utilization:

  a. Industrial and commercial uses of space resources and characteristics.
  b. Potentials for the future.

IV. Political Implications:

  a. Space Law.
  b. International Relations.
  c. Space militarization.

V. Medical Implications:

  a. Physiological aspects of life in extraterrestrial habitats.
  b. Psychological aspects of life in extraterrestrial habitats.

VI. Sociocultural Implications:

  a. Terrestrial:
    1. New opportunities.
    2. Problems that may arise from space humanization.

  b. Extraterrestrial:
    1. Problems that may require sociocultural solutions.
    2. Lifestyle.

3. Relationships with Earth.
4. Government and dependence.
5. Control of deviance.
6. Social class structure.
7. Unique extraterrestrial culture traits.

c. The Future:
1. Space colony or space community?

SPACE: A NEW FRONTIER

Mona Cutolo, Department of Sociology
Denis M. Miranda, Department of Physics
Marymount Manhattan College
New York, NY

ABSTRACT

The challenges and the promises of space colonization present an exciting opportunity for exploring and analyzing the values, the institutions and the physical environments we have created on Earth. Here we describe an interdisciplinary course, team-taught, that examines the current state of space exploration and the innovative technologies spawned by space research. The course also explores the possible social, economic, political and international impacts of migration to space of people and industries. A course project is to design a space colony for a community of 10,000 people. Given the technical design parameters and other details, the students are to engineer socially an ideal community, bearing in mind the short lifetimes of utopian communities of the past. The process is intended to help the students gain a fair understanding of the dynamics of human societies and of the technologies we have developed that enable us to change our world and to design new worlds.

INTRODUCTION

In 1982 at Marymount Manhattan College we offered for the first time a course entitled "Space: A New Frontier." The course was a specific outgrowth of a course on science, technology and society which had been taught for many years with the topic of space and the future as its final session. Both of these interdisciplinary courses have been taught by a faculty team consisting of a physicist and a sociologist.

Encouragement and support for developing the course came from several sources. The World Future Society Conference in Toronto in 1980 reminded us that the future is coming toward us so fast that we no longer have the luxury of leisurely planning for it. Our academic dean, who had attended the conference with us, expressed to the faculty her strong commitment to preparing our students, most of whom are women, for a future in an increasingly technological world. A college Technology Committee was formed to develop further the technological dimensions of the curriculum. The interest and the commitment of two faculty members, one from the natural sciences and the other from the social sciences, each of whom delighted in the academic exchange between their disciplines, led to the development of a course on space.

The course is intended to draw on the liberal arts as the basis for understanding the new technology to be used to plan and design the future. Its objectives are:

- to encourage students to integrate the disciplinary knowledge they have acquired through their study of the liberal arts,

- to analyze critically the advantages and disadvantages of alternative (physical, technical, social, economic, political) systems in a different environment, and

- to design creatively an integrated, functioning society using technologies appropriate for the physical environment of space.

THE COURSE: SPACE -- A NEW FRONTIER

Earth among the Stars

Our course begins with a brief astronomy lesson on the evolution of the universe and the place of Earth and our solar system in one of the spiral arms of the Milky Way Galaxy. We discuss the structure and the composition of stars, planets and galaxies, the conditions for the evolution of life in interstellar space and on planetary surfaces. Finally, we discuss the evolution of life on Earth, and we clearly make the point that in the case of human species cultural evolution has become the dominant environment for human evolution and moral development. Later in the course cultural environmental planning is viewed as an essential component of technological planning.

Changing Perspectives on Space from Earth

Science fiction has provided many of us with our only view of worlds different from our own, so we discuss the changing perspectives on outer space presented in the science fiction literature. Our perceptions of life "out there" have changed from monsters, UFO's, warlike hostile environments, to benevolent, hopeful, understanding communication with alien life as in "Star Trek" and "E.T." We discuss how science fiction has enabled us to view alternative social, political, economic and value systems in fantasy worlds of fiction which we may now be able to create in fact -- if we believe it is wise to do so.

The Space Age

A quick survey of the space age, the only age known to most of our students, enables us to present the historical context of the early discoveries which provide a basis for our steps into space. From the pioneering researches of Tsiolkovsky and Goddard, through the German rocket program of World War II, and the Cold War of the 50's, which provided the political motivations for the space race between the United States and the Soviet Union, we emphasize the social as well as the technological factors in policy decisions. The development of the technologies as well as the accomplishments of the space program are reviewed, especially the Earth satellite and the biomedical technologies which have had such a major impact on the quality of our life. Spectacular developments in communications technology, Earth-resources monitoring and analysis, meteorological-data collection, innovative biomedical sensors and instrumentations, developments in computer technology and a range of commercial spinoffs such as Tang, Teflon, and Velcro are now part of our daily lives. Perhaps the biggest spinoff of all was an unbridled confidence in our technology -- we could do whatever we wanted to do. Believing that, and with our incredible Moon landings becoming routine with no clear future goals, we became complacent. Money for space activities was siphoned off by the unpopular Viet Nam War. We slipped from our peak of accomplishment into the well of failure. The manned space program came to an end until revived by the Space-Shuttle technology and its implications.

But even during the lull between space adventures the impact of the space enterprise on society was being felt, not only in material benefits (the spinoffs) but in other more profound ways as well. Pictures from space showed ONE planet with no boundaries. The artificial nature of political boundaries was deeply etched into our consciousness. A recognition of the fragility of "Spaceship Earth" and our responsibilities to each other as residents of a global village opened our minds to the possibility of envisioning different political, economic and social relationships between global neighbors and even challenged us to reevaluate our responsibilities to the part of the planetary environment entrusted to our care. Our perceptions of ourselves on this planet have changed. Some of our expectations for the future have changed. We have reached a point in history which has given us a rare opportunity to see beyond the routine patterns of the past and to reconsider our options for the future.

Life in Orbit

With this as a fairly familiar background for the students, we
then present the less familiar and more technical aspects of
providing living environments in orbit. We also introduce the
concepts of space industrialization and colonization, as always, from
the perspectives of both technology and social science.

Space labs, space satellites and the Space Shuttle are discussed
in terms of the physics of space travel, orbital dynamics and
rendezvous and in terms of the life sustaining technologies incorporated into the environmental designs. A careful analogy is drawn
between the self-contained life supporting spaceships and our
spaceship, Planet Earth.

Space Industrialization

Space industrialization is introduced by explaining how, in
comparison to conditions on Earth, some of the unique characteristics
of the space environment -- weightlessness, wide range of possible
temperatures, near perfect vacuum, varied electromagnetic radiation,
solar energy -- are superior or preferable for certain types of
industrial activity. Some of the experimental tests performed on
Skylab and on the Space Shuttle have already demonstrated the
advantages of space for some kinds of industrial manufacturing. The
growth of large and pure crystals for the computer industry, the
mixing of alloys, and the development of pure drugs are among the
very attractive industrial applications. The space activity that is
given the highest priority, however, is the harnessing of solar
energy by constructing solar power satellites which could beam
relatively inexpensive, clean (non-polluting), virtually limitless
energy to our energy-starved planet. Successful removal of polluting
energy sources from the Earth and substitution of clean solar power
opens up the possibility of eventually returning the Earth to its
once pristine beauty.

We give the students a lot of technical information in lectures
which are supplemented by modest, in-class demonstrations of some of
the physical principles (artificial gravity, mass driver), by
computer simulations (orbit and rendezvous), and by NASA slides of
space exploration and construction in space. G. Harry Stine's book,
The Space Enterprise[1] provides a good basis for discussion of space
industrialization.

Human Communities in Space

Once the students accept the rationale for industrial development of space, they can begin to deal with the concomitant need to
develop and build space colonies -- a concept which to many lies only

in the realm of science fiction. The first human communities will be
company towns for construction crews. They will probably be constructed using available space hardware such as the huge spent
booster rockets used in Shuttle launches. The construction technologies tested and available, such as beam builders and mass
drivers, are explained as we consider the construction of power
satellites, space factories and human habitats. We discuss the establishment of mining colonies on the Moon and the economic
advantages of mining the nearby asteroids. It is at this point that
several of the complex issues, such as private claims vs. public
rights, individual national vs. international claims to the wealth of
space, are aired. Obviously, some of the issues are intractable: "Do
the nations that have not contributed to the space enterprise have a
claim to the resources of space?" "If only the rich nations can
afford space exploration, how will the underdeveloped countries and
their peoples share in the new wealth?" We refer the students to
the United Nations Law of the Sea as an example of an international
issue with similar geopolitical complexities.

Designing a Colony

Then we begin the major course assignment: TO DESIGN A SPACE
COLONY

Given: The need to develop a space habitat for 10,000 people
by the year 2010.

Purpose: To provide community life for a population engaged in
activities related to mining the Moon and constructing
and maintaining a solar power satellite (SPS) station.

By mid-term each student must select one of the following five areas
and research the issue in the library:

1. Criteria for selection and selection process: who goes into
space, who decides and how?

2. Political structure; governmental system; leadership roles.

3. Economic system within the colony, and its relationship to
Earth.

4. Culture: values and institutions to be adopted.

5. Physical design of space colony with respect to environmental
psychology and architecture.

Students are particularly encouraged to survey the literature on
the frontier experience, utopian communities, urban planning, environmental psychology (especially studies of personnel on arctic
bases, submarines, off-shore drilling rigs, Alaska pipeline), architecture, and human biology. They are required to write a five- to
seven-page paper outlining their recommendations for the space colony
and explaining the basis for them. The class discusses the recom-

mendations on each issue and attempts to come to a consensus on each. Finally, we try to coordinate the recommendations into an integrated social system.

Results and Conclusions

We rarely end the course with an agreed upon, fully designed space habitat. It is very hard to create a new world in three and a half months. But we do bring students to question what they have and what they have taken for granted. We give them reasons to search the past as a tool for creating the future. We introduce them to a range of scientific and technical information which they would not otherwise have sought out.

We provide an issue -- space colonization -- which requires students to integrate the disciplinary perspectives that they have acquired. We think we motivate the lower-division students to explore with a new enthusiasm the liberal arts and sciences as they continue their academic careers. For the upper-division students we provide a theme that enables them to integrate their disciplinary knowledge into a holistic critical analysis of their past and present. We encourage them to explore the vast range of options available to them in making decisions about the nature and the structure of their future world -- be it here on Earth, or at a colony in space.

REFERENCES AND FOOTNOTES

1. G.H. Stine, The Space Enterprise, (Ace Books, New York, 1982).

SPACE COLONIZATION AS A RESOURCE IN THE LIBERAL-ARTS CURRICULUM: THE CASE OF ECONOMICS

Martin Gerhard Giesbrecht
Economics Associates
20 Faculty Place, Wilmington, Ohio 45177

ABSTRACT

Space colonization can contribute to the liberal-arts economics curriculum primarily through the modernization of the paradigms within which the science is practiced. Four examples are listed and briefly described: the re-emphasis on natural resources as a basis for economic productivity; a re-evaluation of property rights necessitated by the increased importance of technology; a re-evaluation of the profit motive also necessitated by technology; and a critique of the metastatic equilibria that constitute economic dynamics today.

INTRODUCTION

As simplified microcosms of "spaceship Earth," space colonies will elucidate and make explicit what is frequently buried beneath complexity on this planet. And, as a prototype of the future, the study of space colonies introduces issues that will have to be dealt with by all future economies, regardless of their location. This is the context in which what follows is to be understood.

IN THE CURRICULUM

Space colonization has not entered the economics curriculum of which I am in charge in any formalized way that might show up in the college catalogue. There are no courses or seminars on space colonization; there have been no workshops for future space inhabitants. Instead, space colonization has invaded the conventional curriculum, from introductory to advanced levels, by way of assigned papers, slide presentations, lectures, discussions, and unprogrammed

intellectual curiosity. It has changed, enhanced, and introduced new ways for students to acquire and exercise their imaginations and their analytical and evaluative skills.

In this paper I will list and describe briefly what I consider to be the four most important influences that the use of the economics of space colonization is having on the curriculum.

Natural resources

Beginning with Marx and continuing through Marshall and Keynes, economic paradigms have tended to ignore the natural resources on which economic productivity is largely based. At colleges and universities concern about them has usually been shunted onto the curricular siding called economic geography, not an academic track on which Nobel prizes, government grants, or even tenured positions are much awarded. The study of space colonization, with its necessary emphasis on resource availability and materials flows and balances, corrects the perspective on this issue.

The dynamic definition of resources is especially appreciated. For example, the proposed conversions of space trash, such as lunar rocks and asteroids, into valuable materials, such as metals and biochemicals, help students understand that resources are created more than they are discovered. They are helped to understand one of the most important determinants of national wealth or poverty. They can appreciate, for example, how the richest economy in the world can exist on the same ground that supported only small tribes of subsistence-level American Indians a few centuries before.

Property

Land, labor, and capital are still presented in the conventional economics curriculum as the holy trinity of economic productivity. A few revisionists have added enterprise or management as categories distinct from labor. Almost universally omitted, however, is the cultural ambience of the economy, which includes most importantly its technological level. Space colonies heighten the appreciation for technology. As almost completely artificial environments, they rely first and foremost on the resource of technology. Coincidentally, all advanced economies, those on this Earth not excluded, do so too.

Technology is distinct from other resources not only in that it functions in the economic process differently from either land, labor, or capital, but also that proprietary claims on it are differently exercised. Land is owned by a landlord. Any infringement on it, such as its removal without consent -- by illegal mining, let's say -- or trespass, may be perceived and may be defended against by force. (The force of law may be preferred, but force it is, in any case.) The same is true for capital, which may be stolen and which can be defended by force. And the same is especially true

for labor, because the owner of labor, the worker, is so immediately connected to the delivery of labor. It can be coerced or stolen by fraud, and it can be safeguarded by force.

But technology is not owned in the same way. It has no significant physical substance. Its theft often goes unnoticed, because the owner perceives no loss. It is difficult to identify ownership of technology by such expedients as serial numbers, labels carrying the owner's name, or even copyrights and patents. It is secured less by force than be secrecy. It can be given away or sold at no significant cost to the original owner. As technology becomes the most important economic resource, the nature of economic proprietorship is certain to change.

Profits

The profit motive is so pervasive in economics that some form of it is even imputed to the motives of government administrators and communist officials. When defined broadly as self-interest, it becomes the universal justification of rational behavior. All other motives, such as altruism, community spiritedness, or love and hate, are deemed to be best understood and best analyzed when they are subsumed as various forms of this broadly defined self-interest.

Yet, if technology can be only imperfectly claimed as property, then the profit to be earned from technology's productivity must also be less than perfectly defendable. What then will be the important motives driving economic productivity in a space colony or other highly technical economy? This question is not asked too soon. It is already a basic question in Silicon Valley, in the pharmaceuticals industry, and everywhere else that technology is the main factor of production.

Equilibria

Economics is shot through with notions of equilibria, from the partial Marshallian market equilibrium of quantities demanded and supplied, to the more general Keynesian equilibrium of aggregate savings and investments, to the grand Walrasian equilibrium of the entire economy. This has been both a boon and a bane to economic science. As a boon, it allows economics to use the language of mathematics, which speaks almost entirely in the syntax of equations. As a bane, it limits economics, which is inherently dynamic, to metastatic analyses, that is, to a dynamics simulated by sequences of static equilibria.

All the flows of resources, productions, consumption, materials and energies in space colonies must be intentionally managed. Do equilibria again play a central part in our understanding and management of these flows? Do we once again freeze them perceptually into sequences of idealized still pictures? Or can we develop a

truly dynamic economics? If the latter, it could unlock the metastatic shackles in which so much theoretical economics is immobilized today and free us from the constraints of equilibria.

CONCLUSION

The most important influence, then, that space colonization can have on economics is to improve the paradigms within which the science is practiced. The above four examples are accessible and meaningful to liberal-arts undergraduates. They are powerful issues. My very brief descriptions of them suffice merely as quick introductions, and there must certainly be many more such issues.

All too frequently, in its quest for rationality under the old paradigms, economics manages to drain all the excitement and passion out of issues. Space colonization introduces a poetry about possibilities that reaches far beyond the usual mundanities of economic science. That is also liberal arts at its best.

CHAPTER 2

IDEAS AND PROPOSALS FOR CURRICULAR USES OF SPACE COLONIZATION

The seven papers of this chapter propose uses of space colonization in the contexts of sociology, anthropology, psychology, philosophy, literary analysis, and economics.

ASKING SOCIOLOGICAL QUESTIONS ABOUT LAGRANGE FIVE

Barry A. Turner
University of Exeter, United Kingdom

ABSTRACT

As human habitation in space seems increasingly feasible, sociologists must begin to identify issues that will emerge. Space colonization raises questions involving the whole range of sociology and offers a novel way of introducing them. Space colonies can be used to explore ideas about communities, particularly utopian communities, and to illustrate important concerns of sociology. One can relate the essential requirements for the survival of any society or community to space colonies, and then go beyond speculation about communities in space and introduce critical analysis of conflicting views in sociological thought.

INTRODUCTION

Let this voluntary confession forestall any future criticism: I am writing about things entirely outside my own experience or anyone else's; things that have no reality whatever. Lucian: Preface to the True History

The idea that men and women might leave the Earth's atmosphere and orbit around the world in space was once a topic suitable only for children's comics, and the thought of men on the Moon was fantastic. Today the idea of a space community with a population of ten thousand people living, working and establishing their homes a quarter of a million miles away from Earth seems, at first hearing, to fall into the same category. And yet, once we have been exposed to the idea, and we start to reflect upon human expansion into space during the past few decades, we realize that we have to think again and treat such proposals as more than mere science fantasy. This realization is confirmed by the recent announcement that an industrial space facility which is to be launched by NASA in 1989 will be joined in the 1990's by a permanently manned space station intended to form a focus for an "industrial park in space."[1]

0094-243X/86/1480056-17$3.00 Copyright 1986 American Institute of Physics

Developments such as these make a space community look technically and politically more feasible, so if we suppose that such a community is to be established, what serious issues do we then need to think about? The generation of students now at our universities will have to live with such developments, and pay for them, so they have a vested interest in taking part in these discussions. Quite apart from such direct concerns, once we start to entertain the possibility seriously, it has its own intellectual attractions. In this novel setting, many of our traditional Earth-bound problems are transmuted so that we start to see them in a new light. How would such a community look? As a sociologist I am interested in considering some of the problems of setting up such a community and in encouraging my students to share some of these intellectual explorations with me (see Appendix).

I have spent a number of years studying large-scale man-made accidents and the social and organizational preconditions of these unforeseen events,[2] and I am very aware that in human affairs there is inevitably a discrepancy between intention and outcome, for, as Mannheim[3] has commented, our "radius of action" exceeds our "radius of foresight." When we act, we generate waves of consequences, intended and unintended, which spread out like ripples on the surface of a pool. Social science may not eliminate all unintended social consequences, for that is an impossible task, but it can try to use past experience and inquiries to point out and avoid some of them.

At some time in the near future an organization like NASA may spend five times as much money as it took to put a couple of men on the Moon in order to build a space community at an Earth-Moon libration point, Lagrange Five. Technologically, this is an incredibly adventurous proposal, but before we proceed too far with it, we need to take some time to think about the unforeseen social complexities of such a project. Gerard O'Neill has been the moving spirit behind this enterprise, and we know from his writings that he wants to explore a new range of technological possibilities for the ultimate benefit of humanity, without thought of pursuing one or another specific solution to the social problems of establishing a space community. He has thus informed us of his "radius of foresight." What then might be his "radius of action"? What consequences, intended and unintended, might such a project set in train?

Can the new community be seen simply as an outpost of our own society, like a scientific station in the Antarctic, or a military base on a Pacific island? A recent fictional treatment which weaves an intrigue of assassination around the framework of the Lagrange-Five project assumes for the sake of dramatic continuity that there is a frequent and easy traffic between the project and Earth,[4] but given the hazard and expense of such a journey, it seems unrealistic to assume that in the foreseeable future such movements will be undertaken for any but the most urgent purposes.

Electronic communications will be readily available, of course, but for all of the concerns of everyday living, it seems reasonable to assume that any space community established at Lagrange Five will have to function as an almost completely self-contained social unit. It is this self-contained quality which poses interesting problems for consideration and forcibly challenges the imagination. It also raises vividly the issues which students of social science should be concerned to explore. Let us discuss a number of these topics as they might pertain to a possible self-sustaining community of ten thousand people established in space at Lagrange Five.

A HEAVENLY PLACE

How far is this community a utopia? In the sense of More's original pun, a space community at Lagrange Five really is a utopia, for at the moment it is a "no-place," a theoretical location in the void, where even the atoms don't congregate very closely together. In the discussions which took place ten years ago at the NASA workshop[5] and in O'Neill's subsequent writings[6] ideas emerged about the kind of community which might be built in space, and although O'Neill explicitly disavows this, inevitably, like all large-scale plans, they have some of the qualities of a dream, or of a perfect blueprint for a portion of the future. He envisages a controlled climate, biological pests eliminated, perfect television reception, industry segregated on a separate satellite, terraced housing with pleasant views where we can serve our friends cocktails and so on. When O'Neill moves on to consider later projects, he offers us an academic haven in heaven complete with "drama clubs, orchestras, lecture series, team sports -- and half-finished books."[7]

Those who look with suspicion on utopian writings stress the un-realistic and ideal features of such abstract societies, often populated by oddly rational beings. They contrast these with the complex, unpredictable world occupied by ordinary everyday non-utopian beings.[8] Some sociologists and philosophers have been par-ticularly critical of utopias.[9] [10] One philosopher, Passmore, urges us to be wary of any schemes which rely for their achievement upon notions that man can be made perfect, since such schemes are often used to justify totalitarian measures which try to correct by coercion the imperfections of those who do not match the blueprint.[11] Passmore's warning is one which I would particularly commend to you.

These reservations, however, do not provide arguments for dispensing with utopias. Utopian writings are extraordinarily diverse, embracing, as Sargent[12] reminds us, models, alternatives, warnings, extrapolations, descriptions and downright fantasy, and they can serve many purposes. Many contemporary utopian writings do seem to turn out as "dystopias" which warn us of the dire conse-quences of various courses of action as we move away from the air of innocence which seemed to accompany many of the nineteenth century attempts to establish ideal communities on Earth. This diversity of utopias can offer benefits. "Man is an animal that can promise," Nietzsche remarks, and utopian writings enable us to examine some of the promises which we are making to ourselves and our children. Polak even argues that our civilization will come to an end unless we continue to produce "images of the future" in utopian writings.[13]

We need, of course, to examine these visions of the future critically, and students who participate in such an exercise should develop judgment about what might or might not be an attainable prescription for the future. Roemer's unique set of course exercises, "Build Your Own Utopia," offers one way of involving students in such an activity.[14]

Utopian writers preoccupy themselves with a variety of possible futures: while some are concerned with the problems of communal living, others want to explore the difficulties and possibilities of feminist utopias,[15] of capitalist utopias[16] or even of the utopias of behavioral psychologists.[17] It has been observed, however, that most utopian writers tend to take a partial view of their posited society, stressing issues which are of passionate interest to them and neglecting to discuss, for example, major economic and political issues, or to give any clear idea of how the young will be brought up.[18] Roemer[19] and Sargent[20] offer extensive listings of utopian writings which may help to guide students who want to embark on a more systematic consideration of the social structure of a future space society.

But for our view of the feasibility of a space community, we do not merely want to know that there exists an extensive and various library of writings which will point to some of the issues, problems and possibilities of "enacted communities."[21] We need also to consider studies which summarize the experience of those who have attempted to set up ideal communities. The earliest stages of a community are of great importance: Stinchcombe[22] has argued that decades or even centuries after their foundation, industrial or-ganizations retain features characteristic of the period in which they were established, and Gherardi and Masiero,[23] in a recent study of co-operatives, contrast the different paths taken and the different characteristics developed by those organizations, on the one hand, which were first established for pragmatic reasons and, on the other, those created by groups of committed idealists.

In the past, many have seen the United States as an arena for trying out their utopias[24] and, indeed, in some respects the American Lagrange-Five initiative shows a continuity with these earlier efforts to improve upon existing society. Attempting to learn from these past experiments, Kanter's major study of American communes and utopian communities[25] points to their very high failure rate, and tries to analyze some of the factors which dif-ferentiate those which continued from those which ended whether with a bang or a whimper.

Kanter invites us to consider the commitment necessary for the establishment of new communities. Those who make a decision to join a novel kind of community must relinquish something in their old life as the price of membership, and establish something in the new one as an investment. The survival of the community through the retention of members depends upon mechanisms of both kinds. Those communities which demand real sacrifice from new recruits are more likely to be successful than those which do not, and those which have distinct and irreversible means of allowing members to make a positive investment in their new way of life also have enhanced chances of survival.

In looking at a space community in these terms, we want to know what the first colonists would give up on Earth to strengthen their motivation to remain members of the community, and what mechanisms of investment would ensure that they committed to the community their individual "profits," however defined, so that they will not readily leave. In view of the difficulties of travel already mentioned, it might be that the decision to abandon Earth and to undertake a risky journey of 250,000 miles would represent sufficient sacrifice, with the hazards and costs of return forcing people to throw in their lot fully with the new community.

On an emotional level, successful communal enterprises require the new member to relinquish links and associations which might disrupt cohesion in the new group. For example, the private, intense relationships occuring between married couples have sometimes been seen as a threat: giving this intimacy up, they can then attain within the group a communion, an experience of collective unity, which binds them to the novel enterprise. Once again, the sheer distance involved in setting up a space community is likely to ensure that early settlers on Lagrange Five will have no option but to sever strong relationships with friends on Earth. However, the ways in which work groups, childbearing and rearing, sexual relations and property ownership are to be arranged within the space community are much more matters of internal organization and ideology. We shall come back to these issues in a moment.

The final dimension of Kanter's analysis concerns the tendency of successful utopian communities to require individuals to undergo some kind of "mortification" which reduces the individual's sense of a "separate, private, uncommited ego" and increases his or her will-ingness to adopt a new identity based upon the "power and meaningful-ness of group membership." As a novice abandons secular names and secular identity upon becoming a full member of a religious order, so utopian recruits are encouraged to become "new persons" as they join unreservedly with the new community. To assist this process, com-munities also develop ways of offering a "transcendence" of the individual ego in a positive form. They promote an ideology of "institutionalized awe" which provides positive and powerful shared beliefs. The current emphasis in Western management upon employees developing "corporate loyalty," a willingness to subordinate individual needs to those of the company's "mission," offers a parallel although usually less intense example of the pursuit of such group commitment through transcendence.[26]

In assessing the possibilities of the space community, we are led to ask whether there will be any such mortification or shedding of the individual's former ego. Will there be a "loyalty screening," and will members be required to demonstrate their commitment to the colony's "mission" before they get their seat on the Space Shuttle? If they are asked to strive for such corporate transcendence, what kind of a community will this create? If they are not, will they threaten division and disruption once they have joined the community?

Or would the processes of emigration to Lagrange Five, willy-nilly, have the effect of severely devaluing previous Earth-based identities, forcing a transcendent commitment to the community, even if such a commitment were not explicitly demanded? And if those of us left on Earth have such a committed, unified group circling around our heads, how will we feel about it?

In looking at the space community as a utopia, then, one focus of my attention, and of that of my students, would be upon the presence or absence of social mechanisms of sacrifice and investment, of renunciation and communion, and of mortification and transcendence. These are desirable to meet the survival needs of idealistic communities established in order to create an existence apart from the world. How far they are likely to be important for a community which is to be established away from the world is a matter for discussion. Will radio and TV links be sufficient to ensure that the distant community remains a part of Earth society in spite of the distances involved, or will the new colonists start to generate their own sense of separateness and difference from the world, as in LeGuin's fictional dystopia, The Dispossessed?[27] Will they feel misunderstood, and make the inevitable fatalities from the hundreds of shuttle journeys into martyrs to embody the sacrifice which they themselves will have made in moving with their futures to an orbit in space? Pioneering journeys such as the Islamic haj, Mao Tse Tung's Long March and the wanderings of the Jews have typically created a strong sense of group loyalty, justified by an ideology which explains the sufferings of the journey, and which separates the voyagers in social and, often, in spiritual terms from those who have not accompanied them. Is this how the "Lagrangians" will come to see themselves?

FUNCTIONAL NEEDS

Turning from utopias, we can look at the needs of the new community more sociologically without assuming that it will be set up with the communal emphasis which is at the center of Kanter's study. We can search for possible unintended consequences of Lagrange Five by making the "thought experiment" of considering, in this new setting, a range of issues which have traditionally been the central concerns of sociology.

We can look at some of the most basic questions about human society, starting with the conditions for its survival. Such issues were considered in the nineteen-forties and fifties in the context of the so-called "functionalist approach" which then prevailed in sociology. The questions raised by functionalist writers must still be considered by students, although, of course, they also have to go on to look at the critical literature generated in response to the functionalists.[28]

The writings of Aberle and his colleagues offer us a useful starting point,[29] defining a society as:

> . . . a group of human beings sharing a self-sufficient system of action which is capable of existing longer than the life-span of an individual, the group being recruited at least in part by the sexual reproduction of its members.

To exist at all, a society has to solve a number of problems. It has to ensure its physical safety, and to provide food, warmth and shelter. To avoid biological extinction it also has to provide the physical conditions necessary for reproduction. Once the grouping can subsist at this level, an array of social matters must be attended to. Externally some means for dealing with other social groupings will be required, and its continuance requires that it is not absorbed totally by another society. Anarchy, a "war of all against all," would have to be avoided, and no society can persist if its members relapse into a completely apathetic state. These last two conditions are simple to specify, but their achievement depends upon establishing a much wider set of social practices, some of which are specified in the "functional prerequisites" which Aberle and his colleagues see as essential for the continued operation of a society.

To function socially, a space community would need to provide means of socializing and rearing the young within the family or some other suitable setting. It will need to have a mode of communication and ways of achieving shared cognitive understandings. Some kind of social goals must be perceived and shared by groups within the society, if not by all members of it, and appropriate means of behaving and of expressing emotion need to be developed. The community is likely to be complex, so that role differentiation and the assignment of individuals to different roles will be called for. Finally, there will be a need for social controls over disruptive behavior. Let us now look in a little more detail at each of these issues.

There are many possible ways of organizing the essential matters of sexual recruitment and provision for biological reproduction. Will children be born and raised in the context of a monogamous marriage, single-parent families, polygamous or polygynous families? What forms of the family are likely to develop? Will there be any pressure to provide communal kibbutz-like nurseries for children? Will the size of the breeding population available allow free mate choice for breeding? Or will such an isolated, limited community have to segregate itself into exogamous clans with breeding permitted only with members of other clans? Anthropological studies of the varieties of kinship systems might suggest some of the constraints here. A possibility which Huxley envisaged in some of his dystopian writings is now readily available to the space community: "DF-AI -- deep freeze artificial insemination." Will this option have special importance in a space town? Lesbian separatists might see Lagrange Five as an ideal spot to use artificial insemination to develop a community without men!

Human babies are not born as social beings. If isolated, they have no genetic mechanism which automatically generates the ability to talk, to relate to others, to become human beings in a social sense. A newborn baby only becomes human through the socialization provided by others, in the context of the family, the nursery or any other social setting in which it finds itself in its formative years. How will these matters be handled for the young in a space community? Will there be kindergartens and schools for the young children of the community? Will there be colleges? Will there be universities, or will all bright youngsters have to make the risky journey to Earth to gain a degree? What are the implications of this? How many towns of this size can support a university?

Any society needs a shared, learned symbolic means of communication between its members. What will this be on the space community? There will be no physical problems about transporting our normal modes of speech, writing and electronic communication, but will the language be Chinese, the language of the majority on Earth? Or French, the language of diplomacy? Swahili, the African lingua franca? Spanish? Arabic? Hebrew? Esperanto? Or could it just possibly be English? Aberle, et al. suggest that in addition to means of communication, a society must establish what they call a "shared cognitive orientation," so that the bulk of the population have a minimal common understanding which will enable them to deal meaningfully with social situations. Will members of a space community have any difficulties in reaching such a common view, or are we to assume that they will already share a common cultural outlook?

Much of the critical comment which has surrounded the functional school has focused upon the next three prerequisites:

- the "need" for a "shared, articulated set of goals" for members of a society (though not necessarily a fully integrated set of goals)

- the "need" for some "normative regulation of means" of expressing emotion

- and regulation of the ways in which it may be expressed.

That is to say, there will have to be some specification of what people should strive for in their actions, and of how they should act appropriately in the society.

The debate about these questions centers upon the extent to which there is or can be any agreement upon such matters in a society, and whether the specification of such prerequisites merely serves to justify the interests and position of those in power in the society at a given time, providing an ideology which makes it easier to eliminate dissent, opposition or even cultural heterogeneity. The corporate view of society, expressed from within an organization, whether industrial, military or bureaucratic, is at home with ex- hortations for the citizenry to act according to an agreed official set of goals, but those of us who are uneasy about the idea of an all-embracing corporate state would point to the claims for dominance built into such views. In the same way, the prescription of "appropriate modes of behavior" raises the question of the sanctions to be applied to those who do not conform. Are those who spend their leisure time listening to twelve-tone music, riding motor cycles or reading science fiction automatically suspect?

These objections do not do away, however, with the prerequisites relating to goals and norms. There is still some need for social agreement if apathy and anarchy are to be avoided, and there is also some need for workable ways of conducting social life, both with regard to cognitive matters and matters of affect and emotion. More recently, sociological writers[30] have looked at social life not as something imposed from above but as an outgrowth of a continuously created enterprise in which all members of a society share. In this view, we would need to turn our attention to considering what kind of social patterns those living and working together in space would create between themselves, and what influences would push them towards one kind of outcome rather than another.

Religion and religious beliefs are a central part of culture, crucially related to the issue of goals and values and what people see as worthwhile in their lives. What form might religion take in a space community? Initially at least, if the experience of the astronauts is any guide, there is likely to be an intensified sense of awe and wonder, and a very strong need to develop an adequate set of personal explanations and beliefs which will help members of the community to adjust to the fact that their life is being lived in an artificially constructed setting in space rather than on Planet Earth. What religions will this be associated with? Will there be mosques, synagogues, chapels or churches on the colony? Are new beliefs and sects likely to develop? Will astrologers still be in business up among the stars?

All societies have social distinctions related to work tasks and to organizational position. How will a space community provide different roles in its social life, and how will individuals be assigned to such roles? Allocating different technical jobs to people usually creates a social division of labor with its own statuses, relationships and tensions. These form the constituent elements in the "social drama of work."[31] What "social drama" will develop around the tasks of Lagrange Five?

Unless an ideology explicitly prohibits it, rewards are distributed within a society in relation to different ceremonial, work or other roles. What kinds of inequalities will exist in this "town in the sky" and what form will they take? Will there be millionaires and paupers? Will there be class differences, will there be status differences, status striving and competition? What will be the medium in which rewards are given? Will there be money? Or company chits? Or communal ownership? Or will all of these issues of reward, status and rank be dealt with simply as a by-product of a military organization occupying the colony? In a military community, there is no ambiguity about the specification of rank, of uniform or of differential rates of pay.

Can we imagine that a society in the sky will provide an environment in which poverty will be abolished? Will the "undeserving poor," those who fall below the absolute poverty level, beg along the walkways of Lagrange Five? Or will a minimum level of social welfare benefits be provided? Presumably, even if absolute poverty is not to be contemplated, relative poverty of the kind discussed by Townsend[32] will still exist. Or will there be no citizens of this new society who will fall to a level where they are incapable of taking a full part in the life of the society around them? Will distinct migrant communities develop with their own ethnic characteristics? Will unsatisfactory migrants be deported?

All societies need some effective control of disruptive forms of behavior. This goes beyond the kinds of social disapproval we have already noted to more active kinds of sanctions. There will be a need, that is to say, in the space town, for laws and regulations, for penalties for law-breaking and for law-enforcement agents of some kind. Will this imply a system of courts, jails and police forces? Or vigilante groups on the "High Frontier"? Or a form of citizen's control, and people's courts? When we establish a space community, we will have orbiting over our heads not only the clear-eyed, square-jawed heroes of the space age, selflessly working for humanity, but also potentially, the whole array of major and petty criminals: space forgers, embezzlers, murderers, muggers, pickpockets, and drug dealers. Is a gun control law an infringement of a citizen's rights on a space colony?

The traditional discussion of "functional prerequisites" raises an important series of questions for students to consider when they try to look at the problems of a space community, or to design their own space community. There are, however, some important questions which are raised only in passing by the "functionalists." These are all issues of coercion, domination, power and control.

Turning again to basic sociological ideas, the state is commonly defined by sociologists as that organization with a monopoly of the legitimate use of the means of violence in a given territory. Allowing that "territory" changes its meaning in space, questions about force and the legitimate control of the use of force enter centrally into political issues in the space community. Who is in control? Will this be a community with a direct line of authority from a military commander? Will it be a company town, with control resting ultimately with a corporation, or with an Earth-based government or international agency? Will it be a small town democracy with political parties and elections, or even a commune? Who exerts the legitimate right to exert violence if someone chooses to challenge them, and how would they do it?

And how do these issues of control and dominance relate to material wealth and material interests in the space community? The community is envisaged as being self-supporting because of its industrial work for Earth and because of its redirection of energy beams to Earth-based power stations. When Earth pays for these goods and services, who will receive the money or other form of reward? Will it be shared equally among the community, or will there be a free enterprise or a planned economy? And if the latter, who plans it? How will work groups be organized and employed, and what will be their social connotations? Will workers join trade unions, or will they be non-union labor? Will the Marxist assumption that social organization, social stratification and social beliefs all mirror the distribution of power associated with control over the means of industrial production prove to be true in the space community? Or are there reasons why this should not be the case?

The technical complexity of the Lagrange-Five project is such that it is difficult to imagine it being planned and executed other than by a strong and powerful central authority, and the most likely prediction that we could make for the future is that such an authority would find it difficult or impossible to relinquish its control over the community's inhabitants and their affairs. Should it be otherwise, however, so that there is a genuine wish to devolve power to the inhabitants and not to run the community as a military or as an industrial command post, then some of the current exercises in encouraging participation in developing countries might offer useful guidelines. Certainly such a devolution would be in line with the spirit of discussions about Lagrange Five to date.

In a recent analysis of the problems of establishing a pattern of participation, Askew[33] argues that several steps are necessary to move a community towards such a practice. Firstly, participation itself will need to be presented and promoted as a value, as a desirable mode of organizing and guiding social life. Appropriate norms of participatory behavior would need to be offered, and individual members of the society would then have to be motivated to behave in ways which fitted in with these norms and values. With such a matrix of social possibilities established, individuals would need to be encouraged to occupy specific social roles where they could display their motivation appropriately. And finally, they would need to be provided with facilities which would enable them to maintain this kind of behavior over time.

There is a dilemma in directing people to participate, as there is in all such projects in developing countries. But if Lagrange FIve is ever to escape the all-pervasive influence of the agency which is set up to establish it, this dilemma will need to be resolved. If it is not, we might at some time see a rerun of the Boston Tea Party at Lagrange Five!

CONCLUSIONS

Asking sociological questions about the kind of community which might be established at Lagrange Five is a fascinating exercise which carries us into all spheres of social life. It brings sharply into focus the whole range of basic sociological issues and offers a novel way of introducing these to students. Sociology students will in all probability want to go beyond this paper to consider many of the questions that the limited space available has forced me to set on one side.

Before we leave the topic, however, there are a couple of concluding points to be made. In presenting such an issue to students, the first question which they will ask is why should anyone contemplate such a project. They might want to ask, for example, whether the enormous amount of money involved might not be spent much more effectively in establishing ideal communities on the fringes of the Sahara to alleviate famine and to help us hold back the growing ravages of the Sahelian drought. Is this the kind of project that only a technologically intoxicated civilization could envisage, or is it a genuine pioneering adventure? Is it an expression of the human potential to voyage into the solar system, or will it become part of a plan to offer the people "bread and circuses" to draw their attention away from more "down-to-earth" political struggles?

Moreover, while the present discussions about space communities have been undertaken wholly in an academic and a civil framework, we have now reached a point where the U.S. Department of Defense is spending more on space technology than NASA, so that it is becoming increasingly difficult to separate civil and military issues in space. When the Moon can be discussed as a "nice military base," as a potential "battleship," and when it is apparently current orthodoxy that "Space is now the high ground of military technology," as U.S. Major General John Storrie recently remarked, it is difficult to see Lagrange Five being used wholly for peaceful purposes. Artists often display an accuracy in striking to the heart of a matter, and it is clear to anyone watching Star Wars films, or reading science fiction stories such as The Lagrangists[34] that the use of outer space for military purposes will definitely be on the agenda unless serious steps are taken to prevent this happening.

If we can avoid exporting terrestrial military preoccupations into space, and if the depredations created on the high frontier can avoid some of the excesses associated with frontier life here on Earth,[35] we may look forward within the next four or five decades to the presentation by a sociologist of an empirical community study of Lagrange Five.

REFERENCES

1. "A Place for US Industry." London Times, August 21, 1985.
2. B.A. Turner, Man-Made Disasters, (Wykeham Press, London, 1978).
3. K. Mannheim, Man and Society in an Age of Reconstruction: Studies in Modern Social Structure, (Kegan Paul, Trench, Trubner, London, 1940).
4. M. Reynolds, The Lagrangists, (Tom Doherty, New York, 1983).
5. Space Settlements: A Design Study, NASA SP-413, edited by R. Johnson and C. Holbrow, (U.S. Government Printing Office, Washington, D.C., 1977).
6. G.K. O'Neill, The High Frontier: Human Colonies in Space, (Cape, London, 1977); G.K. O'Neill, 2081: A Hopeful View of the Human Future, (Simon & Schuster, New York, 1981).

7. O'Neill, The High Frontier, p. 227.

8. E. Weber, "The Anti-Utopias of the Twentieth Century," South Atlantic Quarterly, Summer, (1959).

9. K. Popper, The Open Society and Its Enemies, 2 vols., (Routledge & Kegan Paul, London, 1945).

10. R. Dahrendorf, "Out of Utopia: Toward a Reorientation of Sociological Analysis," Am. J. Sociology, 64, 115-127, (1958).

11. J. Passmore, The Perfectibility of Man, (Duckworth, London, 1970).

12. L.T. Sargent, "Authority and Utopia: Utopianism in Political Thought," Polity, 14, (4), Summer, 565-584, (1982).

13. F. Polak, The Image of the Future, 2 vols., translated by E. Boulding, (Oceana, New York, 1961).

14. America as Utopia, edited by K. M. Roemer, (Burt Franklin & Co., New York, 1981)

15. B. C. Quissell, "The new world that Eve made: feminist utopias written by nineteenth century women," in Roemer, ed., op. cit., pp. 148-174.

16. L.T. Sargent, "Capitalist Eutopias in America," in Roemer, ed., op. cit., pp. 195-205.

17. B. F. Skinner, Walden Two, (Macmillan, New York, 1948).

18. L.T. Sargent, "Utopia and Dystopia in Contemporary Science Fiction," The Futurist, 6, (3), June, 93-98, (1972).

19. Roemer, ed., op. cit.; K. M. Roemer, Build Your Own Utopia: An Interdisciplinary Course in Utopian Speculation, (University Press of America, Lanham, MD, 1981).

20. L. T. Sargent, "Utopia and Dystopia in Contemporary Science Fiction," The Futurist, 6, (3), June, 93-98, (1972); L.T. Sargent, British and American Utopian Literature, 1516-1975: An Annotated Bibliography, (G.K.Hall, Boston, 1979).

21. G.F. Sutton, "Structural Features of an Enacted Community," Communal Societies, 4, 49-58, (1984).

22. A. L. Stinchcombe, "Social Structure and Organization" in Handbook of Organization, edited by J. G. March, (Rand McNally, New York, 1965), pp. 142-169.

23. S. Gherardi and A. Masiero, "The Impact of Organizational Culture on Life-Cycle and Decision Making Processes in Newborn Cooperatives," Dipartimento di Politica Sociale, University of Trento, Italy, 1985, unpublished.

24. Roemer, ed., op. cit.

25. R.M. Kanter, Commitment and Community: Communes and Utopias in Sociological Perspective, (Harvard University Press, Cambridge, Mass., 1972).

26. B. A. Turner, "Sociological Aspects of Organisational Symbolism," Note-work: Newsletter of the Standing Conference on Organisational Symbolism, 4, (2), 9-17, Spring (1985).

27. U.K. LeGuin. The Dispossessed, (Gollancz, London, 1974).

28. N. J. Demerath, III and R.A. Peterson, System, Change and Conflict: A Reader on Contemporary Sociological Theory and the Debate on Functionalism, (Free Press, New York, 1967).

29. D. F. Aberle, A. K. Cohen, A. K. Davis, M. J. Levy, Jr. & F. X. Sutton, "The Functional Prerequisites of a Society," Ethics, 60, January, 110–111, (1950).
30. H. Garfinkel, Studies in Ethnomethodology, (Prentice Hall, New Jersey, 1967).
31. E. Hughes, Men and Their Work, (Free Press, Glencoe, Ill., 1958).
32. P. Townsend, Poverty in the United Kingdom, (Penguin, Harmondsworth, 1979).
33. I. Askew, "Community Participation in Health and Family Planning Programmes: An Organisational Perspective," unpublished doctoral thesis, University of Exeter, Exeter, UK, 1984.
34. M. Reynolds, op. cit.
35. T. Kroeber, Ishi: A Biography of the Last Wild Indian in North America, (Univ. of California Press, Berkeley, 1961).

APPENDIX

TEACHING HANDOUTS: Introduction to Sociology 1985/86,
University of Exeter

LECTURES

11. Introductory: Why should anyone want to establish a space
community? What is the technical background? What kind of society
do they envisage? What problems do we see?

12. Prerequisites of a self-sustaining community: Can we specify
what is necessary for a community to operate at all? Do the writings
of the functionalists help us here? Can we apply the literature on
functional prerequisites or does this need modifying? How relevant
are criticisms of the literature?

13. Theoretical discussions of community: Do we get any help from
theoretical writings on community and society? Does the work of
Tonnies, Durkheim, Marx help us to understand the nature of a self-
sustaining community, society or social system?

14. Clues from community studies: Communities have been studied by
sociologists and anthropologists in a variety of (earthbound)
settings. What can we take from these empirical studies to help us
to specify the nature of our community in space, or to warn about the
dangers of trying to establish a community in space?

15. Utopias, dystopias and communes: Are those discussing space
colonies looking at utopias? When we review writings on eutopias,
utopias and dystopias, does this give us a better understanding of
the proposals to establish such communities? There have been many
attempts to establish ideal communities on Earth as communes, as
religious settlements or as other social experiments. Are studies of
these attempts relevant to forecasts about space colonies? Do they
offer blueprints for success or warnings of failure? What progress
have we made in understanding the nature of Community?

TOPIC: COMMUNITY

Background

In this first extended topic of the course we shall look at
"community," an idea which has been of interest to sociologists for a
long time. We will range fairly widely over topics related to this
idea, but the central focus for our work will be plans currently
under discussion for the possible establishment of space colonies of
10,000 people in space-orbit around the Earth.

From the perspective of this course, the technical aspects of these developments are not important. A NASA Summer Workshop in 1975 and a follow-up conference in 1985 regarded the idea as technically feasible using today's technology. It is possible to construct and operate self-sustaining space colonies, orbiting the Earth at a distance of one quarter of a million miles, a level much higher than current space satellites and well above the orbits at which President Reagan is now talking about trying to establish his "Strategic Defense Initiatives." Such colonies would rotate to provide artificial gravity, would obtain energy from the Sun to power their machines and would grow their own food. At first they would justify their existence economically by manufacturing activities and by building reflectors to concentrate solar energy for use as a power source on Earth. If you are interested in this, look at Johnson and Holbrow's Space Settlements.

You will see from Johnson and Holbrow, from the writings of O'Neill and from the first lecture that the main thrust in thinking about such colonies has been scientific and technological. O'Neill explicitly denies that he wishes to be involved in planning any particular form of society in space. He is not trying to establish a utopia but to explore with his physics students a technical possibility which could increase the freedom of action of the human race by moving it away from the surface of the planet. He suggests that establishing the first community would take six to ten years and cost from three to five times as much as the Apollo project which put men on the Moon.

Essay

Against this background, I would like you to imagine what kind of social arrangements might be possible in such an orbiting entity. We will look at a number of aspects of this problem in the five lectures, considering the idea of community, and the constraints, requirements and possibilities of people living together. You should write for your tutor an essay on one of the topics set out below, each of which requires you to explore different aspects of the sociological requirements if a viable community is to be established and to maintain itself. The setting which we are considering is one in space orbit, but the conclusions you draw might have implications for tribal communities, for towns and villages in the United Kingdom, or for the "utopian" communities which people have set up in their minds or in the real world.

You should write an essay of 8–12 pages on one of the topics below. Some references are given in the course literature, but you are also encouraged to explore the library to find more relevant material. Make sure that you understand how the library catalogue works, and use the subject index to locate your own relevant material.

The International Encyclopedia of the Social Sciences is another useful source. Your aim is to tackle the essay topic from a sociological point of view. To understand how a sociologist might look at these problems, the course texts will be useful. Also read

Peter Berger's <u>Invitation</u> <u>to</u> <u>Sociology</u>, or the new paperback <u>The</u>
<u>Sociology</u> <u>Game</u> by Anderson, Hughes and Sharrock (Longman, 1985,
  3.95). Looking at present or back issues of journals such as the
<u>British</u> <u>Journal</u> <u>of</u> <u>Sociology</u>, the <u>American</u> <u>Journal</u> <u>of</u> <u>Sociology</u> or
other journals shelved at 300-305 will extend your thinking about
your essay. You will also find the literature on utopian communities
worth exploring for some of the essay topics.

<u>Topics</u>

1. Discuss the important sociological elements which would need to
be considered if a self-sustaining community of 10,000 people were to
be established in a space orbit.

2. Outline in <u>sociological terms</u> the characteristics of a utopia
that you would like to see established in a space community;

OR

Outline <u>in sociological terms</u> the characteristics of a dystopia that
you would like to warn society against establishing as a space
community.

3. Imagine that you are a sociologist living in a space community in
the year 2035, carrying out a study of the social system surrounding
you. Outline the main features which you would have to consider in
your study, linking these to studies carried out before 1985.

4. Show how sociological discussions of communities, from Tonnies
Redfield and Wirth onwards, could be brought to bear on the problems
of a space community.

5. Present a carefully argued critique of proposals to establish
space communities, outlining how you would see such proposals
relating to current concerns in contemporary society.

6. <u>With the prior approval of your tutor</u>, you may write an essay on
some other topic which deals with sociological issues raised by
attempts to establish a social community in space.

SPACE COLONIZATION AS A TOOL FOR TEACHING ANTHROPOLOGY

Thomas L. Melchionne and Steven L. Rosen
Department of Anthropology
Rutgers University, New Brunswick, NJ 08903

ABSTRACT

One hundred years of anthropological research has sought to discover the properties of human nature. This research bears directly on the problem of creating new societies in alien environments. Space colonization presents theoretical and practical problems which anthropology can help solve. These problems and the attempt to solve them can be used in the classroom as a vehicle for teaching both ethnology and physical anthropology. In such a course students would explore the findings of both cultural and biosocial anthropology, and use these findings to construct a space colony which has reasonable prognosis for survival.

INTRODUCTION

Greg Barr,[1] the administrator of the L5 Society, pointed out that "space development is the obvious interdisciplinary subject that ties together all the fields of human endeavor."

This paper shows that the kinds of questions which the social engineers of a space colony will have to answer are precisely those that anthropologists have been grappling with for the past century. For this reason anthropology has the potential to make great contributions to the field of space development. Some anthropologists have already started to work in this field.[2],[3]

This paper illustrates how the problems of space planning can be used to teach anthropology at the college level. It does this by showing

1. how the various subfields of anthropology are relevant to space development, and

2. how considering the problems of space development can be used as an heuristic device to teach these various subfields

An anthropology course on space colonization will have students synthesize cross-cultural and biosocial data in order to plan a prototype space colony. In doing this the anthropology student will first examine the vast corpus of cross-cultural and biosocial data already in existence. This will enable her to learn how the various terrestrial societies have coped with the problems of life. Utilizing what she has learned, she will then decide which terrestrial patterns, if any, would be suitable for a space community.

BIOSOCIAL ANTHROPOLOGY AND SPACE

Humans are Primates

We are still primates. It is hubris to pretend that we are not. Genetically _Homo_ _sapiens_ is as closely related to the great apes as the chimpanzee and the gorilla are related to each other. The DNA codes in the blood proteins of chimpanzees and humans are ninety-nine percent identical. Five million years ago, just yesterday in evolutionary terms, the ancestors of _Homo_ _sapiens_ _sapiens_ and _Pan_ _troglodytes_ (the common chimpanzee) diverged into separate species.[4]

As humans strive to become like gods and leave the planet, our mammalian primate nature will reassert itself. As Desmond Morris pointed out,[5] despite the cultural and technological achievements of the human race, man "is still very much a primate . . . Even a space ape must urinate." Our bodies and brains are still those of our hunting-gathering-scavenging ancestors who evolved on the African savannah. It is their hormones that are secreted by our glands, and their blood which courses through our veins.

The Power and Limits of Culture

This is not to deny that culture alters the conditions of human life to an astonishing degree. Culture is our species's greatest adaptive tool. It has enabled us to thrive on every continent, speculate about God, cook Peking Duck, and write woodwind quintets. The very fact that we are seriously discussing the colonization of space testifies to the power of culture. No dog, no rat, no serpent, no ape, ever sought to gain an off-planet foothold for his species. Because of culture, _Homo_ _sapiens_ is the first terrestrial species to have the settlement of space within its grasp.

Significant as it is, however, culture's power to alter the circumstances of human existence is not unlimited. People seem predisposed to live in certain sociocultural arrangements, and not in others. For example, men are permitted to have more than one wife (polygyny) in 75% of the Earth's cultures; monogamy is the rule in 24%; polyandry, the practice of women having multiple husbands, is found in only 1%; while group marriage is almost nonexistent.[6] It should be pointed out here that monogamy is as common as it is, because it prevails in the large number of world societies which have adopted Christianity. It thus seems that biology, and not culture alone, limits the possible range of marriage forms. If randomness rather than biosocial predisposition determined how unrelated cultures developed their marriage practices, one would expect a more even world distribution of the four different possible marriage forms than 75%, 24%, 1%, and nearly 0%.

The relative frequency of the four different kinds of marriage is only one example. The anthropological literature is full of data which indicate that certain forms of social organization occur with greater frequency than others. The preponderance of data of this type suggests that despite the versatility of the human animal, Homo sapiens is not infinitely malleable. Some of the limits to the versatility of the species are the result of simple biological constraints. For example, humans have primarily primate locomotor and sensory potentials. Other limits are the result of terrestrial ecological conditions, such as the need for clothing and warmth.

Childbearing in Space

Even in space, certain aspects of the human condition will not change. Women, not men, will still bear the children. How will this affect division of labor in the settlement? Anthropological research on Israeli kibbutzim has shown that even in those intentional communities which adhere to the ideology that women and men should have equal access to all professions, the overwhelming amount of childcare is provided by the women; the women choose to be the childcare providers.[7] Is it reasonable to believe that the situation in space will be any different?

Homo sapiens is characterized by an extraordinarily long period of childhood dependency. Children will still have to be fed, clothed, and educated. Who will do this? Only the biological parents, the community at large, or a combination of both?

Adolescence in Space

Like other mammals, human males are potentially sexually active from puberty onwards. But unlike other mammals, human females do not experience oestrus and are thus also potentially perennially sexually active. This biological reality presents problems for both the student of anthropology and the planner of space settlements.

At what age and in what manner will adolescent sexuality be allowed to be expressed? In a space colony, the emphasis on technical and scientific education will necessarily be great.

Therefore, the education process will be long. This being the case, young people will not be economically productive members of the society until long after reaching sexual and biological maturity. One problem for the planner of space settlements will be how to negotiate the bestowal of adult sexual and reproductive rights on people who are physically and emotionally adults, but who are not economically ready for parenthood.

Space colonies will have to develop social mechanisms to deal with the reality of their young people's sexual drives. If a colony's population is small, where will adolescent colonists find partners of their own age? Will the community tolerate age disparities in sexual partners, sexual relations among blood kin, and/or adolescent homosexualty?

Marriage in Space

If the institution of marriage is to exist in some form in the society, where will young adults find marriage partners? Will there be an extra-terrestrial equivalent of computer-matched mail-order brides and husbands? During the 19th century British colonials returned home on leave in order to find marriage partners. Would a shortage of potential spouses require people born in the colony to return to Earth for the same reason? Finding suitable marriage partners will particularly be a problem if the colonies are to be permanent settlements. It would not be a problem in the case of outposts populated by short-term contract workers who arrive at the colony as a husband-wife team and plan to depart after a few years.

Will the rules regulating marriage be generated and enforced by some governmental authority? Will religion have a role in determining the nature of marriage? Is it possible for a society to flourish with no rules to regulate relations between the sexes?

The Aged in Space

Even in space people will still get old and die. Will the very old be shipped back to Earth or will there be facilities to care for them? If people are to spend the bulk of their adult lives in space, it might be reasonable to assume that they will not want to return to Earth in their old age. Indeed, they might not be physically capable of doing so; who will care for the aged?

Ben Bova[8] has suggested that, "it may well be that the first people to live permanently in space will be the old, the weak, the infirm." This is because the low gravity will enable old and weak people to remain active and productive in ways which would be impossible on Earth. Regardless of how long low gravity enables vigor to be extended, there will come a time when infirm individuals will simply become too weak to be active, even in low or zero gravity. This presents the problem of how the space colony will care for such individuals. Invalids who have physiologically adjusted to their condition by living at low gravity can not be expected to survive terrestrial conditions. They will not be able to be returned to the Earth for their final convalescence and will have to be cared for in space. The possibility that space colonies may become vast nursing homes for these formerly active old people is a social issue that both space planners and anthropology students will have to address.

CULTURAL ANTHROPOLOGY AND SPACE

Designing a Society

In planning a space colony, anthropology students will be asked to design a system of social organization for the colony ab initio, using as their database the corpus of ethnographic literature. Anthropology has catalogued hundreds of different forms of social organization. Planners of intentional communities in space will have the entirety of the ethnographic and historical record at their disposal to draw upon as a database. Students working on social organization will compare the various possible kinship arrangements, principles of descent, inheritance, and marriage forms which have existed in terrestrial societies. They will analyze this information and then decide which kinds of social organization would be most suitable for life in the proposed settlement.

This process will have the advantage of enabling students to view the social organization of their own culture in a relativistic light. It will allow them to see how other social systems might be more appropriate in certain other circumstances. For example, students will be forced to confront the issue of whether or not polygny or polyandry might be preferable to monogamy in some circumstances; whether political authority should be invested in the entire population, in a select group, or in a single individual; whether the legal basis of the society should favor the rights of the group or the rights of the individual. These questions will be raised, because there is no good reason to expect that the social organization of intentional communities in space will mirror that existing in contemporary North America. This will be especially true if space colonization is to be an international multi-ethnic venture.

Culture Shapes Personality

Historically, psychological anthropology has been chiefly concerned with determining the nature of the relationship between culture and personality. Anthropological research shows that different forms of sociocultural organization reward, punish, and thus give rise to different personality types. Depending on the sociocultural character of the proposed space colony, different psychological profiles may be desired of the inhabitants. Psychologists will have an important role to play in determining the kinds of people who will staff future space settlements, just as they have been involved with the crew selection process for Antarctic research and for manned space endeavors.[9],[10] Psychological anthropologists, specializing as they do on the relationship of personality to culture, can be of value in determining the kinds of individuals who will be most suited for life in space. In a course on the anthropology of space, students will consider the extent to which individual personalities are shaped by the social institutions, and the applicability of this to space issues.

Other questions will also be addressed besides those concerning population selection. Will institutions be designed to maintain and legitimize predetermined conceptual and behavioral orientations? To what degree is it possible to engineer personality types? To what extent is it ethically and/or pragmatically desirable to do so?

Law in a Space Colony

Law and jurisprudence is another area in which anthropology may be of help. Ever since 1861 when Sir Henry Maine wrote his treatise on Ancient Law, anthropologists have been compiling information on how primitive and ancient societies have resolved disputes arising among their members. What kind of legal system should exist in the space colony? How will disputes be settled? How will criminal offenders be punished? On the one hand, space colonies will be highly technologized and sophisticated societies. On the other hand, they will also be, at least at first, small-scale societies with at most a few thousand members. This being the case, will the legal systems of nation-states with populations numbering in the millions be appropriate? Or will a system of law characteristic of small face-to-face primitive societies be more desirable?

How Will Resources be Shared?

Will there be private property, or will everything be held in common? Will the colonies have money, or will everything be free? Will the creation of individual fortunes be permitted, or should everyone in the society live at the same socioeconomic level? Will people be permitted to own property? Will the means of production be controlled by financial interests on the Earth or only by people who live in the colony? Who will benefit from the profits? What types of arrangements do the ethnographic and historic record show are possible or desirable and in what circumstances?

Illness in a Space Colony

The medical anthropologist is concerned with the cultural context of illness. Such contexts influence how illnesses are defined, treated, and experienced by the members of a culture.[11] This is no less true of illness in a contemporary urban hospital, than it is in a Tiv village in the African bush. It will also be true in space.

Not only physical illness but mental illness is culturally determined.[12] This being the case, it is to be expected that new kinds of mental illness will erupt in space settlements.[13] In order to understand the etiology and the pathogenesis of these new mental illnesses and to design correspondingly appropriate treatments, the mental health practitioner must be aware of these new space-specific cultural determinants. In assignments, anthropology students can speculate on how life in space will radically alter the kinds of diseases which will be encountered as well as the cultural context of the healing process.

Relations to the Environment

Issues of human ecology will also be addressed. Homo sapiens's relationship to the environment will be utterly changed. Only those plants and animals that are intentionally brought to the colony will be part of the settlement's ecosystem. Human control over the environment will be greater than ever before. The character of this control will be an artifact of planning. Should only domestic animals which are 'useful' to humans be brought along, or should there be an attempt to synthesize as much as possible a semi-wild environment with animals such as foxes? Is it psychosocially healthy for humans to live in a world where all the plants and animals exist solely to serve people?

Religion

Traditional religious beliefs will also be called into question. At least since Frazer's Golden Bough[14] anthropology has examined the ways in which religious and magical beliefs structure human conceptual and social universes. Humans are not only Homo faber (man the maker) but most fundamentally Homo religiosus.[15] Religion is concerned with establishing, for each person and for each society, the ground and goal of existence.[16] Even in a nonterrestrial environment humans will have to confront the problem of ultimate meaning. To what extent will terrestrially originated religious systems still be tenable? Is the creation of a sacred/religious Weltanschauung necessary or desirable? Is there a role for the social scientist here, or must the colonists await for the coming of a religious genius? In the attempt to answer these questions, the student will be asked to examine the ethnographic record to see how religious myths, magic and rituals have served to create and maintain social orders throughout history and across cultures.

Life Transitions

The student will be asked to examine how rites of passage enable people to pass through various life transitions. In any permanent space colony, people will be forced to confront the same existential crises that they have always had to face on Earth. Rites of passage concretely reenact cosmologies on the human plane. Birth, death, puberty, marriage, menopause, all these life-cycle events must be endured and meaningfully constructed.[17],[18]

The student will also be forced to consider how shamanistic and sacerdotal technologies,[19] (e.g. trance states, meditation, body disciplines, song, psychotropic and hallucinogenic drugs, ecstatic dance), can be used not only as escape valves for the stress of life in space, but also for the expansion and transformation of consciousness.[20] These archaic techniques of ecstasy[21] when used in the denatured environment of a space colony might enable the colonists to forge new myths. Paul Ricoeur[22] has argued that modern man can no longer believe in the old myths but must "de-mythologize" them in order to arrive at their essential meanings. If this will be true on Earth, how much more so in space? The student of the anthropology of space, with access to the multitude of world mythic systems, can distill essential meanings relevant for the new life in the space colonies of the future.

CONCLUSION

Now in the late 20th Century, voyaging among the planets and colonizing the void with the seed of humanity is not an idle fantasy. The technology, if not the financial backing, already exists for such an endeavor.[23],[24] It is only a matter of time before it happens.

Homo sapiens is an endangered species, and its tenure on the Earth has not been a long one.[25],[26] It has been pointed out, both in the SF literature and in the serious research on the subject, that the colonization of space is a way to ensure humanity's existence, regardless of what happens on Earth.

Anthropology is particularly well suited to the application of space issues. It operates at the interstices of the territories claimed by other fields and so offers to the space-policy planner a way to address the key questions of the human condition, just as space issues provide an excellent medium for the teaching of anthropology.

REFERENCES AND FOOTNOTES

1. Greg Barr, "From the High Ground: Introspection," L5 News, 10(8):0, (1985).
2. Cultures Beyond the Earth, edited by Magoroh Maruyama and Arthur Harking, (Vintage Books, New York, 1975)
3. Interstellar Migration and the Human Experience, edited by Ben R. Finney and Eric M. Jones, (University of California, Berkeley, CA, 1985).
4. S.L. Washburn and Ruth Moore, Ape Into Human: A Study of Human Evolution, 2nd ed., (Little, Brown and Company, Boston, 1980).
5. Desmond Morris, The Naked Ape: A Zoologist's Study of the Human Animal, (Jonathon Cape, London, 1967), p. 23.
6. George P. Murdock, "World Ethnographic Sample," American Anthropologist, 59(4), 686, (1957).
7. L. Tiger and R. Fox, The Imperial Animal, (Holt, Rinehart and Winston, New York, 1971).
8. Ben Bova, The High Road, (Pocket Books, New York, 1981), p.189.
9. Robert L. Helmreich, John A. Wilhelm and Thomas E. Runge, "Psychological Considerations in Future Space Missions," in Human Factors of Outer Space Production: AAAS 50, edited by T. Stephen Cheston and David L. Winter, (Westview Press, Boulder, CO, 1980), pp. 1-23.
10. Kirmach Natani, "Future Directions for Selecting Personnel," ibid., pp. 25-63.
11. George M. Foster and Barbara G. Anderson, Medical Anthropology, (Wiley, New York, 1978).
12. V. Crapanzano and V. Garrison, Case Studies in Spirit Possession, (Wiley, New York, 1977).
13. Patricia Santy, "The Journey Out and In: Psychiatry and Space Exploration," The American Journal of Psychiatry, 140(5), 519-527, (1983).
14. James Frazer, The Golden Bough: A Study in Magic and Religion, (Macmillan, New York, 1945, 1891).
15. Mircea Eliade, Shamanism: From Primitives to Zen, (Harper and Row, New York, 1974, 1951).
16. John E. Smith, Experience and God, (Oxford University Press, New York, 1968), p. 15.
17. Peter Berger, The Social Construction of Reality: A Treatise in the Sociology of Knowledge, (Anchor Books, New York, 1966).
18. Victor Turner, The Ritual Process: Structure and Anti-Structure, (Aldine Publishing Company, Chicago, 1979).
19. Jerome Rothenberg, Technicians of the Sacred, (Doubleday and Co., New York, 1968).
20. William Irwin Thompson, Passages About Earth: An Exploration of the New Planetary Culture, (Harper and Row, New York, 1973).
21. Mircea Eliade, op. cit.
22. Paul Ricoeur, The Symbolism of Evil, (Beacon, Boston, 1979).
23. Gerard K. O'Neill, The High Frontier: Human Colonies in Space, (Morrow, New York, 1977).

82

24. Ben Bova, op. cit.
25. L. Tiger and R. Fox, op. cit.
26. Nigel Calder, "Insurance to Save Our Species," Science Digest,
    February 85(2), 13-17, (1979).

PSYCHOLOGY AND SPACE: FROM THE FRYING PAN INTO THE FIRE?

Fay Terris Friedman, D'Youville College,
320 Porter Avenue, Buffalo, NY 14201

ABSTRACT

The needs of space colonization pose many questions relevant to psychology. What would be the psychological consequences of living in space? What personality characteristics would you wish in your companion colonists? How would you design a space colony to avoid undesirable psychological consequences? Are we ready to go? The answers to these and other questions could form an interesting psychology course.

PROBLEMS OF GOING

I would start the course by having students imagine what it would be like to live in a space colony.

- what would they gain -- psychologically, materially?

- what might they lose -- psychologically, materially?

- what would they want to take along with them and how feasible would this be? Could they learn to do without things like record players, books, drugs, if they had to? What substitutes could be provided?

To make them aware of some of the technological problems of a space flight I would use an exercise called "NASA Exercise Worksheet"[1] in which students must assess from a list of items, e.g., two 100-lb tanks of oxygen, food concentrate, etc. (15 items), which would be most essential in an emergency where the space ship was forced to land at a point distant from the rendezvous point on the Moon. Students have to rank order these in terms of their importance to the crew. This could make them aware of conditions which are different in space such as weightlessness and the lack of air.

WHO WOULD YOU PICK TO GO?

Who would you select to go into space? It will be important to choose persons who can cope with the extraordinary conditions as well as those who can work best and would be most compatible with others. This would be a fine opportunity to apply personality theory to a real(?) situation. Each student would be asked to choose a crew of 12 people on the basis of the following types and justify the choice:

- Conformist -- nonconformist

- Introvert -- Extrovert

- Internal locus of control -- External locus of control

- Endomorph, mesomorph, ectomorph

- Rational (formal thinking) -- Intuitional (works on hunches)

You could discuss the disadvantages as well as the advantages of specific choices. The same exercise could be done with occupational groups.

MAKING SPACE LIVEABLE

Here you could introduce the concept that space psychologists like Yvonne Clearwater hold dear: that psychological needs are more critical than technological considerations in space living. Or are they interactive? Do you have to know a great deal about human needs before you can engineer a space craft for living?

At this place Clearwater's article "The Human Place in Outer Space: A New Challenge for NASA Planners -- Creating a Space Station Environment where People can Live and Work Well for Months or Years" could be assigned.[2] According to Clearwater, space living will start as early as 1992. We could then discuss what she sees as possible psychological problems in space living -- restlessness, depression and boredom -- and have students come up with plans to offset these possible conditions.

This is a good point to discuss research on living in isolation such as studies of submarine living and arctic isolation, as well as accounts of living in space such as Cooper's A House in Space.[3] Some of the questions to be considered would be:

- What differences in behavioral priorities exist under conditions of physical strain and deprivation?

- How does behavior change under conditions of weightlessness? (An opportunity to consider the meaning of Abraham Maslow's Hierarchy of Human Needs.)

- Would it be advisable to use drugs to alter the nervous system to make life more comfortable, or would stimuli such as music accomplish the same results?

- What happens to an individual's sense of identity and uniqueness in a closed group? Does group identification offer an adequate substitute for possible loss of individual identity?

The problem of intimacy, which Dr. Clearwater does not neglect in her article, could be broached. The whole question of sex as well as gender could be considered along with the benefits and possible disadvantages of having both men and women in space. Clearwater says,

> We have noticed that in training and study, the whole work atmosphere and the mood in a crew of men and women are better than in men-only groups. Somehow the women elevate relationships in a small team, and this helps to stimulate its capacity for work.[4]

Clearwater also believes that a normal healthy crew will have normal sex appetites and that planning for "intimate behavior," i.e., for auditory and visual privacy, must be a consideration in planning for space living.

Students will be encouraged to submit original thoughts on the subject.

AN OVERALL PICTURE

At this point I would like students to get an over-all picture of life in a large space settlement as Gerard O'Neill conceives it in The High Frontier[5] or Heppenheimer in Colonies in Space.[6] Social and sociological aspects not yet mentioned are brought up here.

I would assign readings of articles in the collection Nova: Adventures in Science.[7] The section on exploration has two appropriate articles outlining space ventures to date. I especially like the introduction to this section:

> What animal would be so reckless as to venture into a totally different habitat? Taking such a risk would seem to defy the whole point of evolution which is the progressive refinement of strategies for dealing with specific environments. . . . Yet humans -- and scientist humans in particular -- persist in ignoring the logic of this reasoning. . . . Our aims: to see what (or who) if anything exists in the hinterlands of our universe, to learn more about how our own home was created by studying

other planets, and to search for new ways to help life on
Earth. All very sound -- and true reasons. But one
can't help wondering whether a simpler motivation
applies; we feel compelled to reach for the stars
. . . because they are there.[8]

Do students agree with this statement? Or would they tend to
agree with a psychologist who said "Most of us couldn't care less if
the universe began with a big bang or a little whimper, if the Moon
really is made of green cheese or even if the Earth is flat."[9]

This should arouse some interesting discussion.

To keep the course going I plan to use as a textbook Mary
Connors' (and associates') fine NASA publication called Living Aloft:
Human Requirements for Extended Spaceflight,[10] published in 1985.
It happens to be an especially well-written book dealing with such
human problems in space as crowding, boredom, emotional stability of
residents, the effects of rejection in a small group, the "long eye"
and sleep disturbances, the importance and meaning of work, verbal
and non-verbal communication, and so on. It also contains a bibliography of approximately 800 books and articles on the subject.

This book should give the student an appreciation of, not only
the problems of living in space but on Earth as well. It should
alert them to aspects of living not often considered, for example,
the effects of gravitation, of atmospheric pressure, and of temperature changes on our physical and psychological well-being.

Students would write an original paper on a topic such as:

- Designing a home in space with human needs as a primary consideration. What kind of research would be necessary before
  planning such a residence?

- Psychological disorders likely to be generated by living in
  space.

- Selection and training for people in space. Do we want to
  avoid elitism?

- Will there be greater expectations for cooperation in space, or
  more likelihood of war?

- Those interested in medical and physiological psychology could
  consider how physical needs in space impinge on psychological
  well-being, e.g., sleep, food, medical attention, possibility
  of outbreak of disease, of panic.

Other possibilities for livening up the course might be

- A trip to Toronto's new Tour of the Universe at the the CN
  Tower.

- Have students read a work of science fiction, and/or write
  their own behavioral science-fiction work.

- Invite guests from the local state university, SUNY Buffalo, or
  from Niagara University, or the L-5 Society.

- Have students attend an L-5 Society meeting.

CONCLUDING REMARK

Somewhere Ray Bradbury has said that the selves we put in space must be better than the selves we are here. He says there's not much use in traveling to other worlds if we can't do better than we've done on Earth.

I say, "Sorry, Mr. Bradbury, we can't wait that long; we can't wait around a few millennia until we can evolve a better species of humanity." We are going to have to send into space a fallible species of Homo sapiens and hope that somehow conditions in space will provide a near utopia where people will learn to live together somewhat more peacefully and productively. And it is with that in mind that I will book one of the first flights to Island One.

REFERENCES AND FOOTNOTES

1. J. Johnson, Instructional Strategies and Curriculum Units for Secondary Behavioral Sciences, (State University of New York, Plattsburgh, NY, 1973), pp. 29-32.
2. Y. Clearwater, "A Human Place in Outer Space," Psychology Today 19 (7), 34-43 (July 1985)
3. H.S.F. Cooper, Jr., A House in Space, (Holt, Rinehart and Winston, New york, 1976).
4. H.S.F. Cooper, Jr., ibid., p. 43.
5. G.K. O'Neill, The High Frontier, (Doubleday, New York, 1982).
6. T.A. Heppenheimer, Colonies in Space, (Stackpole Books, Harrisburg, PA, 1977).
7. Nova: Adventures in Science, (Addison-Wesley, Boston, 1983).
8. Nova, ibid., p. 83.
9. B. Zilbergeld, The Shrinking of America: Myths of Psychological Change, (Little, Brown and Co., Boston, 1983), p. 135.
10. M.M. Connors, Living Aloft: Human Requirements for Extended Spaceflight, (NASA Scientific and Information Branch, Washington, D.C., 1985).

LIMITATION AND LIFE IN SPACE

Marvin Israel and T. Scott Smith
Dickinson College, Carlisle, PA 17013

ABSTRACT

"The Earth is the very quintessence of the human condition . . . ," says Hannah Arendt. Georg Simmel writes: "The stranger is by nature no 'owner of soil' -- soil not only in the physical, but also in the figurative sense of a life-substance which is fixed, if not in a point in space, at least in an ideal point of social environment." How will no longer being Earthbound affect persons' experience of themselves and of others? Space colonization offers an opportunity for new self-definition by the alteration of existing limits. Thus "limitation" is a useful concept for exploring the physical, social and psychological significance of the colonization of space. Will people seek the security of routine, of convention, of hierarchy as in the military model governing our present-day astronauts? Or will they seek to maximize the freedom inherent in extraordinary living conditions -- as bohemians, deviants, travelers?

LIBERAL ARTS, SOCIOLOGY AND SPACE

Sociology as a liberal art can foster self-knowledge by articulating assumptions shared with others and which underlie our thinking and actions. Such sociology is interdisciplinary because its raw material comes from diverse sources ranging from physics to literary criticism of science fiction. The literature on space colonization encompasses almost every field of endeavor imaginable, and students who analyze this literature and their responses to it deeply enough cannot fail to learn something essential about themselves. The inexperienced student, however, needs some assistance both in finding resources in the writings of sociological theorists for analysis and in seeing their relevance and how they might be applied.

Sociological theory deals with the problem of how social order is possible. This approach lends itself readily to the educational goal of self-knowledge described above. One approach, called Analysis,[1] attempts to answer that question through the articulation of the beliefs upon which all ordering activity is based. The question of how a particular social order is possible can be answered by articulating the belief system which underlies it.

WHY LIVE IN SPACE?

Students should start by reading the apostles of space colonization and listing the justifications given for colonizing space:

- There is the hedonistic justification; think here of the "artists' conceptions" resembling vacation condo resorts[2] and the forested comets envisioned by Dyson[3] -- a kind of Club Med in space.

- There is the capitalistic justification; think of Heinlein's The Man Who Sold the Moon -- space as the last bastion of in-dividualistic striving for monetary gain -- a kind of Andrew Carnegie meets Ayn Rand.[4]

- Similar, but not identical, is the fortune hunter who will hate space but endure it for the high salary like his counterpart working on the Alaskan pipeline.

- There is the justification of social responsibility -- the visionary who in the long run will relieve Earth of her social problems in contrast to the short-sighted idealist who wants to do something about poverty right now by expending resources on Earth.

- There is the justification of providing another frontier for the adventurer to challenge, thereby providing a heroic model for those who stay behind.

- There is the justification of providing an escape to those facing hopeless conditions -- the Statue of Liberty in space.

Then there are the arguments of those who are opposed to space colonization, for example, government figures like Senator Proxmire who seem slow to fund solar power satellites, the likely first step to colonization.

LIMITS ON EARTH AND IN SPACE

In guiding students to reflect on some of the issues associated with space colonization it is pedagogically useful to have them explore the role that limits and limitation play. It is especially useful to understand how people decide what constitutes a limit and what stance they adopt toward limits. While, to our knowledge, these ideas are not addressed directly by sociological theorists, they are nonetheless implied in their writings.[5]

Students can analyze the arguments of potential space dwellers and their opponents in terms of limits where a limit is any kind of scarcity. In these terms the arguments mentioned above advocate escaping limits whether we are talking about escaping the scarcity of pleasure or the scarcity of energy.

Often such arguments don't consider the possibility that what one person perceives as a limit, another perceives as an open horizon. For example, let us look at the often heard claim that the shortage of energy on Earth coupled with the availability of an almost infinite supply of solar energy in space, which could be collected and beamed back to Earth via microwaves, compels us to embrace this technology. When asked to think about it, students can see that there is something problematic about the idea of scarcity of energy. This is not just an issue for hard science but also for the social sciences and humanities. Considering the environmental problems connected with the widespread use of automobiles for personal transportation in the Los Angeles metropolitan area, some might argue that too much energy is available there. Similarly, the neighbors of Three Mile Island might feel that too much energy is concentrated in one place, given the threat of nuclear accidents. Some conservationists and environmentalists, coming to the problem of scarcity from the humanities, believe that we have more energy now than is good for us because it has served to encourage gluttony and sloth.

The effect of environmental features on our behavior can be understood only by determining what significance we have assigned to them. These effects are the result of an interaction between the physical "things" and our thinking about them. Energy is not just the number of kilocalories available in material. An infant contains usable energy just like a piece of wood, but how many parents in danger of dying of exposure would perceive their child as a possible source of fuel? (A point similar to this is made in the film "Soylent Green" in which dead bodies are processed for food, but that fact must be concealed from consumers.) Is the energy that we as individuals have for doing our work not partly a function of how interesting that work is; that is, does not the interaction between our desires and the task have something to do with the "fact" of how much energy is available?

Students should be made aware of the ironic possibility that in seeking to escape the limitations of Earth, they might create a social environment with equal or worse limitations. There are many historical examples of severe social limitations imposed in the name of achieving freedom. The Puritans left England for the New World to gain religious liberty and then constructed a highly restrictive society. Most political revolutions at some stage oppress in the cause of liberty. Students could use the concept of limitation when considering what form the social organization of space colonies would take. They could use it to examine the two major forms that have been mentioned in the literature, namely, the rigidly organized, militaristic or bureaucratic form and the bohemian, free-spirited, loosely organized form.

Again it is important to keep in mind that limits have benefits. Emile Durkheim, one of the founders of sociology, argued that limitation is the foundation of idealism and service to humanity. Freedom from limits does not allow the individual to develop to full capacity, instead it leads to a deformation of the human spirit.[6] Here Durkheim is working with the same model as Raffel[7] -- a model of human nature traceable to the Ancient Greeks -- in which human nature can be understood as always in a state of seeking or desiring. Limits preserve desire by compelling us to test them in order to find out what our limits actually are. This process ends only at death.

To get students to realize that limits result from the interaction of self with what is "out there," it helps to deal with phenomena that are closer to their everyday experiences and concerns than energy shortages. To stimulate reflection on the values underlying the decision to stay on Earth or to colonize space, especially justifying colonization as a way of escaping shortages (limits to growth), ask students what they might do about burdensome or undesirable conditions in their own lives, anything from the ethnic or religious background they were born with to the roommate assigned to them at college.

Students often complain that there is a scarcity of desirable members of the opposite sex at their school. What do they think should be their response to this; should they seek to escape this perceived limit or should they submit to it? (Submission in our usage also includes rebellion, since this still treats the condition as an unescapable fate.) Upon reflection, the student may see that to submit (that is, to conform or to rebel) to a condition as though it imposed an unambiguous limit upon one's action is tantamount to alienating oneself from one's physical and social environment, so that one loses sight of one's own responsibility in construing the meaning and relevance of what exists in that environment.

"Submission" to this perceived scarcity could involve such behaviors as competing for the small number of "desirable" partners or retreating from the social arena into complaining or work. Escape could involve weekend trips to other schools to find partners or even transferring to another school. Both of these alternatives treat scarcity as "factual."

But what could scarcity mean here? A serious relationship always involves artificially creating a kind of scarcity for oneself in that the kind of action appropriate to a situation of abundance would be tantamount to betraying one's commitment to one's partner. That is, one voluntarily limits one's resources to one other person of the opposite sex. It is possible that a perception of scarcity is expressive of a kind of greediness and promiscuity. Then, too, the perception of too few desirable partners may be dependent upon a conventional and somewhat impoverished notion of desirability, for example, that a person be resourceful at partying. What are the criteria used to choose potential partners? If the occasions of socializing could be expanded to permit different criteria -- the ability to engage in intense and focused conversation rather than small talk, for example, or the capacity to enjoy a walk rather than dancing -- desirable partners formerly believed scarce might now be seen as abundant.

Let us, in a similar manner, consider how students might be led to reflect on the realities of traveling to a new place.[8] Travel is always looked forward to in a haze of anticipatory pleasure. Descriptions and artists' renderings of space colonies in O'Neill's book[9] and in Space Settlements[10] resemble the promotional brochures for vacation properties -- everything is neat, pastoral, and orderly. As in advertisements for suburban homes, the people are portrayed as comfortingly conventional and moderately affluent. Some small proportion of travel does turn out something like our fantasy of it, but the rest is an ordeal of arranging the minutiae of daily life in an unknown environment where our taken-for-granted expectations are constantly being violated by alien ways. The most recent of countless such renderings of the reality versus the fantasy of travel is the movie "European Vacation." Here the unflagging enthusiasm of the father of the family, who has won a trip to Europe, is contrasted with the boredom of viewing the official sights, the distastefulness of unfamiliar food, the contempt of the natives, the sheer incomprehensibility of much of what one is suffering.

One irony of travel is that the trip, initially seen as an opportunity to be free of the restrictions of our ordinary life, is often turned by the traveler into something of a straitjacket. It is as though there were something in us which demanded limitation and discipline, leading us to develop sightseeing schedules as rigid as the work schedules of astronauts.

ARE EARTH'S LIMITS AN ESSENTIAL PART OF BEING HUMAN?

One can hardly escape the impression that some of those recommending space travel are so concerned to normalize it that they write about it as though it were only technically rather than essentially different from any other trip. Even though the justification most often cited for space colonization is to escape Earth's limitations, it seems that the importance of Earth's limitations is not acknowledged as seriously as it deserves to be. The major social thinker Hannah Arendt[11] says: "The earth is the very quintessence of the human condition . . . ." She explores the significance of only one aspect of this insight, namely, how the literal escape from Earth in space flight is actually preceded by the much earlier figurative escape from an earthbound perspective in the adoption of modern scientific thought, and how treating scientific thought as a paradigm for all thought deprives us of certain uniquely human qualities. She addresses certain unforeseen and unfortunate consequences resulting from modern science's escape from the limitations of pre-scientific thought. In doing so, she leads the reader to question the value of always trying to escape limitations rather than working within them.

Although we will not reproduce the complex turns of thought by which she develops her characterization of the modern age, her work could be valuable in encouraging students to think seriously about the issues involved in space colonization, especially with respect to the attitude one adopts toward the limitations of Planet Earth.

To say that the Earth is the quintessence of the human condition is to insist that it is important to remember that human beings are originally earthbound. To be earthbound means to have a body. It is by now a commonplace insight in medicine that human beings consist of both body and mind, but doctors have not often explored the significance of this insight, nor, except in a technical sense, have the proponents of space colonization. One doctor, a well-known neurologist, is an exception to this generalization. In the aftermath of a serious mountaineering accident which resulted in the total disappearance of nerve transmission to his leg, he was stimulated to reflect on the mind-body link.[12] He found that his inability to sense his leg -- its weight and strength -- produced the feeling that the leg was not his, that it was an alien object. This was such a profoundly disturbing experience that it affected his perception of reality per se.

How much more our sense of reality would be affected if, as the result of a weightless environment, our entire body were to feel unfamiliar! Heppenheimer[13] has a short discussion of "sex in zero g," but he confines himself to mentioning the increased opportunity to experiment with different positions without seriously addressing the social and emotional implications of this opportunity. A guided discussion in the classroom of sex in zero g could elicit some thoughts on the significance to a woman of the man's weight on her body during the sex act. The film "Quest for Fire" portrayed the male superior position as the crucial step in the transition from beast to human. A number of feminists have argued that the pressure of the male's body on the woman is an icon of their socially hierarchical relationship, that to achieve equality between the sexes requires addressing much of the subliminal body language present in heterosexual transactions. A weightless environment reduces the significance of the male's greater muscular strength which even among upper-class families, researchers have found, profoundly influences cross-sex relationships.[14] One of our earliest and most profound experiences as earthbound creatures is the sense of efficaciousness we get from overcoming resistance by exerting our strength. What would be the effect of losing this sense in a weightless environment?

It might be argued that we are able to produce gravity artificially by spinning the structures in which people would be living in space. But (and here we return to Arendt) the very knowledge that it is our will which decides whether or not there will be gravity (or how much) -- the knowledge that what we formerly took to be a parameter of our experienced reality is really relative to our inclinations and therefore arbitrary -- makes our experience of gravity-related phenomena essentially meaningless. In other words, when deprived of the absolutes derived from our earthbound vision, human nature cannot flourish.

Human nature as conceived by such major sociological theorists as Weber, Durkheim, Simmel and Arendt, follows a life course guided by value judgments. Value judgments depend upon a belief in an unchanging standard, and the typical life as conceived by these theorists pursues self-improvement measured against absolute ideals. Weber showed how the importation of modern science's abstract mode of thought into human organizations led to the substitution of what he called "formal rationality" for "substantive rationality." That is, organizational behavior could be consistent and logical but completely crazy from an ethical point of view. Simmel[15] showed how the stranger's "objectivity," made possible by the total absence of emotional ties between a person and members of the group in which one is a stranger, produced the emotionally impoverished life of the permanently alienated being. In space we are all strangers.

These thinkers conclude that human beings need limits in the form of absolute standards. They see life as a process of simultaneously striving to grasp what the limits are and to live up to them. Students should be asked to think about what forms of ethical behavior are in some sense dependent upon those features of Earth's environment which are not yet under our technological control -- not just gravity, but also the length of the day, the seasons, air pressure, and all such similar matters. Will space colonizers in an attempt to develop substitutes for the absolutes which were left behind on the Earth impose tyrannical, rigid regimes on their new homes? Will the solar system be divided -- as in Dyson's scenario -- with the inner solar system area of abundant energy and scarce water the province of bureaucracy and with the comets of abundant water and chancier energy becoming the frontier of the rugged individualist?

CONCLUSIONS

It may sound as though we are opposed to space colonization, but that would be to misunderstand our position. Whether it colonizes space or stays on Earth, humanity still has to face the question of how best to cope with its conditions and limits. By promoting relectiveness on our students' part, we can hope to insure that the attitude they take to space colonization will have been developed on the basis of thoughtful self-knowledge.

Limits are important in many contexts, and students need to understand their nature and impact. Since many proponents of colonizing space see it as a way to escape the economic, political, or social limits of Earth, space colonization is a useful context in which to discuss limits. Sociology provides a useful methodology for the discussion.

We conclude that students should examine the extent to which limits are consequences of our perceptions and consider the necessity of limits and their possible benefits. In this connection several questions are especially interesting. If the limits of Earth are an essential part of our humanity, what will we become in space? How might the limits of space shape the social and political forms of a colony? Are these new forms likely to be freer than those left behind on Earth? Will space colonists want to be freer than they were on Earth?

REFERENCES AND FOOTNOTES

1. Alan Blum and Peter McHugh, Self-Reflection in the Arts and Sciences, (Humanities Press, Atlantic Highlands, NJ, 1984).
2. Space Settlements: A Design Study, NASA SP-413, edited by Richard D. Johnson and Charles Holbrow (U.S. Government Printing Office, Washington, D.C., 1977).
3. Freeman J. Dyson, "The World, The Flesh and the Devil", in Communication With Extraterrestrial Intelligence, edited by C. Sagan, (MIT Press, Cambridge, MA, 1973), pp. 371-389.
4. H. Bruce Franklin, Robert A. Heinlein: America as Science Fiction, (Oxford University Press, New York, 1980).
5. An exception is the work of sociologists working within the Analysis perspective. See, for example, Peter McHugh, Stanley Raffel, Daniel Foss, and Alan F. Blume, On the Beginning of Social Inquiry, (Routledge and Kegan Paul, London and Boston, 1974).
6. Emile Durkheim, Moral Education: A Study in the Theory and Application of the Sociology of Education, translated by Everett K. Wilson and Herman Schnurr, edited by Everett K. Wilson, (Free Press, New York, [1925], 1973), p. 40.
7. Stanley Raffel, "Bob Dylan" in Friends, Enemies, and Strangers: Theorizing in Art, Science, and Everyday Life, edited by A. Blum and P. McHugh, (Ablex, Norwood, N.J., 1979), pp. 19-28.
8. An example of serious sociological reflection on travel (unrelated to ours) can be found in McHugh et al., op. cit.
9. Gerard K. O'Neill, The High Frontier: Human Colonies in Space, (Morrow, New York, 1977).
10. Space Settlements: A Design Study, op. cit.
11. Hannah Arendt, The Human Condition, (U. of Chicago Press, Chicago, 1958), p.2.
12. Oliver Sacks, A Leg to Stand On, (Summit Books, Simon and Schuster, New York, 1984).
13. Thomas A. Heppenheimer, Towards Distant Suns, (Fawcett Columbine, New York, 1979), p. 193.
14. Violence in the Family, edited by Suzanne Steinmentz and Murray Strauss, (Harper & Row, New York, 1974).
15. Georg Simmel, "The Stranger" in The Sociology of Georg Simmel, translated and edited by K. Wolff, (Free Press, Glencoe, Illinois), pp. 402-408.

TECHNOLOGICAL POSSIBILITY AND PUBLIC POLICY

Steven Lee and Scott Brophy
Hobart and William Smith Colleges
Geneva, NY 14456

ABSTRACT

Space colonies are easily imagined in a variety of social forms
that can be used to highlight questions of social philosophy.
Because of their exceptional dependence on technology they are particularly suitable for investigating philosophical questions
pertaining to the development and use of technology and reliance on
it. Space colonies also serve as models in terms of which to explore
broader questions of social justice and the ethical dimensions of
public policy issues. For example, we ask "Is it morally justified
for a government to embark on a policy of rapidly colonizing space
given the resources this would require?" We conclude that the actual
benefits are likely to be much less than anticipated. Regardless of
the possible benefits, a public policy for devoting present resources
to house many people in space is a policy of not devoting such
massive resources to other more pressing problems confronting this
generation of humanity. The idea should not be adopted as policy.

INTRODUCTION

This conference explores the relationship between space
colonization and the liberal-arts curriculum. Our contribution to it
will be an examination of some public policy issues involved in the
development of new technologies, particularly those relating to space
colonization. We will begin by looking briefly at possible uses of
the idea of space colonization in teaching philosophy, and then turn
to an extended example of one such use in the area of teaching
philosophical methods of analyzing public-policy issues. We will
offer an example of such an analysis in the form of an argument
against the moral justifiability of government allocation of large-
scale funding for the development of colonies in space.

0094-243X/86/1480096-9$3.00 Copyright 1986 American Institute of Physics

PHILOSOPHY, SOCIAL JUSTICE, PUBLIC POLICY
AND SPACE COLONIZATION

The academic discipline of philosophy had a good part of its origins in questions about an individual's relationship to social and political decision-making, and it continues to be a goal of the liberal arts to educate individuals to analyze and to participate in a responsible way in matters of public debate. Philosophical analysis of social issues contributes to the aims of a liberal arts education in two different ways. First, in exploring the foundations of social justice, it leads individuals to widen their imaginative vision about fundamental matters and to clarify their thought about some very basic moral concepts. Second, in exploring the ethical dimension of current public policy issues, philosophical analysis can contribute to the educational goal of preparing individuals for informed participation in the formation of public policy.

Developments in technology are of interest to philosophers in this regard. Exploration of foundational issues is often facilitated by thought experiments which involve devices that in one way or another test our intuitions about the adequacy of certain principles. Evil demons, rings that when worn render their wearer invisible, brain transplants, and other such fictional devices have often served to engage the imagination in philosophical thought. For example, there are questions in social philosophy that are traditionally addressed by imagining the features of an ideal society or a society that is starting from scratch. How will such communities govern themselves? What would constitute legitimate forms of moral and political authority in such communities? How could we construct a way of life free from the social evils? Answers to such questions would help us understand the nature of social justice.

In the sixteenth century, the idea of the "New World" played an inspirational role in re-thinking the foundations of social concepts, and centuries later the idea of the "new frontier" provided a landscape for utopian thought-experiments. So the use of such examples is not new. But the idea of a space colony is an especially valuable example of this kind. For when such examples are not only logically possible but technologically possible as well, the thought experiment is usually more readily accessible to the student. When the idea of a new society is thought of as something more than what can only be imagined, when it is thought of as a real possibility on the technological horizon, as space colonization now is, then such an idea becomes especially useful in teaching aspects of philosophical analysis.

Technology may also be of interest to philosophers in the context of public policy questions about allocation of resources in the development of emerging technologies. Resource distribution is not simply a subject for economists and professional policy-makers; there are moral questions at stake, questions about fairness and social justice. Given that public resources are finite, decisions which favor the spending of tax dollars in one way are _ipso facto_ decisions against spending tax dollars in other ways. While it is a reliable principle in ethics that "ought implies can" (that if something ought to be done, it can be done), the converse proposition is ludicrous -- not everything that can be done ought to be done. Some areas of technological development are, for reasons which include those having to do with the fair distribution of resources, simply not worth pursuing.

Principles for Deciding to Develop a Technology

Prominent illustrations of difficult moral choices that emerging technologies often generate may be found in the field of medical ethics. The National Institutes of Health and other government-supported research funding agencies obviously have a finite amount of resources with which they can encourage the development of medical technology. This means that difficult choices must be made, and principles are needed for such decision-making. Suppose, for example, that a certain life-saving device is extraordinarily expensive, and though it may confer great benefit to some, their number would be fairly small. What principles bear on the question of whether such a technology should be developed instead of others which might offer a smaller amount of benefit to a broader range of recipients?

We need to sort out two components of such choices: first, how much benefit may be expected from a technology; and second, upon whom would such benefit be conferred? The following may illustrate how the second of these considerations, as well as the first, is morally relevant. Imagine that two life-saving devices would each contribute one hundred units of utility (however one chooses to assign value to units of utility). But imagine that one of these devices would bestow all one hundred units of utility on one person, and the other would distribute that utility more evenly among the population. Other things being equal, the second device is morally preferable to the first. The question of the distribution of benefit is a morally relevant consideration distinct from that of the total amount of benefit conferred.

Many would argue that it is simply unfair to spend large sums of public funds on technologies that promise to benefit only a few. Some suggest that an example of this in medicine is the artificial heart. Many argue that we ought not to devote the huge amounts of health-care resources that would be needed in order to perfect this device, for the reason that doing so would result in extending the lives of only a relatively small number of people at great public expense. The money would be better spent on other health-care needs.

The constraints of fairness on the distribution of resources are more obvious when the resources in question are public funds to be spent by governments. We will therefore restrict our analysis of the question about the moral justifiability of space colonization to the public-policy case. It may be that in the case of the artificial heart, the current trend toward private funding and development is justifiable, whereas large-scale public funding is not. In the case of space colonization the same may be true. The colonization of space is a matter of public policy insofar as governments are currently the only ones with enough resources to initiate a project aimed at quickly establishing large human colonies in space. If such a project could be financed privately, then moral questions no doubt arise, but they are questions different from the one we will address in what follows. In addition, we will not consider the moral objection to a public policy of space colonization (and much else), raised by advocates of the "night-watchman" state, that tax money should not be spent for any function other than that of minimal police protection of rights to non-interference.

Space Colonization as an Example

Our question, then, may be posed in this way: is it morally justifiable for a government to embark on a policy of rapidly colonizing space given the resources that this would require? There are two quick answers to this question that we should begin by rejecting. First, some might argue that it is not morally justifiable to embark on such a policy because it is not natural for humans to live in space. As a kind of parody of natural law theory such an argument would assert that humans should not (in a moral sense) act in a way contrary to their natural functioning, and that such functioning requires as a necessary condition the natural environment of the Earth's surface. Even if the former assertion is accepted, the latter cannot be -- there is no reason why the environment in which it is natural for humans to function cannot itself be artificially created, as it is, not only in space but in large office buildings and shopping centers.

The second argument we must reject is the one that would conclude that space colonization is not only justifiable, but morally obligatory, based on the assertions that we have a moral obligation to fulfill our destiny, and that the destiny of our species lies in life among the stars. Both of these assertions are highly questionable, but even if we were to accept the first, it is not clear what could ever count as conclusive support for the truth of the second. The rejection of these two arguments shows that there is no way to demonstrate either the moral justifiability or the moral unjustifiability of space colonization as a matter of public policy. A more careful examination of the issue is required.

What should be the basis of a moral evaluation of matters of public policy? Based on our earlier distinction between amount and distribution of benefit, we can say, roughly speaking, that if a public policy is to be morally justifiable, it must maximize social benefit and must not violate considerations of justice and fairness. Let us begin by assuming that a public policy of space colonization would not violate considerations of justice and fairness. The question of its moral justifiability reduces, then, to its defensibility on grounds of social benefit. Space colonization is morally justifiable if and only if it promotes social benefit better than alternative policies. To be more precise, space colonization is morally justifiable if and only if members of the human race as a whole would be in the long run better off were the necessary resources put into such a program than they would be were these resources put to some other public-policy use.

Several points should be made about this formulation of the question. First, the notion of "better off," i.e., the problem of what exactly is to count as social benefit, will be left unspecified, since we believe that a rough understanding of this idea is sufficient for the argument. Second, some might maintain that the class of persons whose welfare is considered should not be the world population, but rather the population of the nation for whom space colonization would be a public policy. We do not believe that the specification of the class makes an important difference in the argument, so we will not take issue with those who wish to understand the case in terms of the smaller class. Third, the argument concerning this question, like any argument based on social benefit, will depend crucially on certain empirical assumptions about what would happen should certain policies be adopted. We hope that such assumptions can be kept clearly in view as the examination proceeds.

There is, first of all, a prima facie case that space colonization is not defensible on social-benefit grounds, similar to the argument suggested earlier against the artificial heart: such a policy would consume a very large amount of resources, material wealth and human talent, that could otherwise be used to improve the general welfare to a much greater extent. This case seems, on the surface, unassailable. It is clear that great resources would be required, and it is equally clear that such resources could be used to great benefit in areas of human services. What is necessary to defeat this prima facie case is, obviously, a successful argument that space colonization itself would provide sufficient benefit to outweigh the loss in welfare from the resources needed to implement the policy.

Here is where the empirical assumptions become crucial. For the sake of simplicity, we may identify in two different scenarios two broad sets of empirical assumptions about the likely outcomes from initiating a policy of space colonization. We may label these the optimistic scenario and the pessimistic scenario.

The optimistic scenario holds out the vision of a rich and abundant future resulting from space colonization. One example of such a vision is presented by Allan Russell,[1] and our discussion will make use of his speculations on the future of space colonization. First, he argues that a very large number of persons could be living in space within a relatively short period of time. Russell maintains that by the year 2200 there could be 1800 million persons living in space. The communities in which these persons live will be small by Earth standards (a few hundred thousand), and each will be an autonomous nation, independent not only of other space colonies, but also of the nations of the Earth (and so will not really be colonies at all). In addition, the space colonies will constitute a tremendous diversity of cultural, social and political forms. Because of this diversity of political forms, and because the communities will be multiplying, by the year 2200, at the rate of one hundred per year, they will provide us with a "science of beginnings," a kind of experimental political theory. Moreover, the lives of the persons in space will be not only rich in a cultural, social, and political sense; they will be rich materially as well.

The pessimistic scenario sees things very differently. The main point of difference with the optimistic scenario would lie in the assumption about the autonomy of the space colonies from the nations of the Earth. The pessimistic scenario is based on the assumption of nonautonomy. It sees the space communities literally as colonies, politically part of the nations of the Earth.

From this point of difference between the scenarios would flow other crucial differences. First, under the assumption that the communities in space are political extensions of Earth nations, it is much more unlikely that there will be living there the large number of persons that is foreseen by the optimistic scenario. This would be the case for several reasons. For one, the colonies are not likely to be regarded as places where people live their lives; they are not homes, but rather places where Americans or Russians go to work, as they now go to work at their nation's bases in Antarctica. Thus, reproduction is unlikely to occur there, so there would be no natural increase in the space population. Another reason that the number of persons in space will remain small under the pessimistic assumption of nonautonomy is that the number will be a function of the perceived needs of the Earth nations that own the colonies, and it is hard to imagine that the purposes of the Earth nations would be served by this number being very large. Another difference flowing from the pessimistic assumption of nonautonomy is that there would be no more cultural, social, and political diversity among the space colonies than there is among the Earth nations, since the former are mere extensions of the latter.

Which should we accept, the autonomy assumption of the optimistic scenario or the nonautonomy assumption of the pessimistic scenario? There are strong reasons favoring the nonautonomy assumption. First, the Earth nations will have a very strong motivation to regard the communities in space as theirs and, consequently, a strong incentive to resist with force any attempts on the part of the colonies to achieve independence. The Earth nations will have spent vast amounts of their own resources to establish the space colonies, and would do so only were they to perceive a great enough national interest in having the colonies to justify that expense. This would create the motivation, both natural and strong, to regard the colonies as theirs, and to defend such claims with force. Of course it is true that after many years of ownership and exploitation, Western nations have in some cases voluntarily decolonized many parts of the Third World. But this is a poor analogy on which to base expectations of voluntary decolonization of space communities. Colonial powers subjugated existing societies, whereas the space communities would literally be created by the colonizing nations; the moral case for decolonization is much stronger in the former case than in the latter.

Second, independence efforts on the part of the space colonies are both unlikely to be initiated, and, if initiated, unlikely to succeed. If the inhabitants of the space communities are only working there and not conceiving of the communities as where they live, the very idea of establishing the communities as independent political entities is not likely to make much sense to the inhabitants or even to occur to them. In addition, the vulnerability of the space communities to military attack would certainly make the threat of force by the Earth nations sufficient to deter any would-be rebels.

What are the implications of these two scenarios for the success or failure of the prima facie case against the moral justifiability of a public policy of space colonization? If the optimistic scenario is the correct one, then a good argument could be made that the prima facie case would fail. Under the assumption that the members of the species whose benefit is of moral interest includes not only those alive today, but in addition those who will be alive in the future, the high level of welfare of the very large number of persons who would be living in space in 2200, under the optimistic scenario, would certainly outweigh the loss in welfare of people alive in the present due to the diversion of social resources to space colonization.

But we have argued that it is the pessimistic scenario that is more likely to be correct. Under this scenario it is clear that the prima facie case stands, if for no other reason than that any higher level of welfare accruing to the persons living in space under this scenario would not be sufficient, given their small number, to outweigh the loss in welfare appealed to in the prima facie case. But there are other reasons. Under the pessimistic scenario, the loss in welfare is likely to go beyond that brought about by the diversion of resources to the project. For one of the main purposes for which nations would build space communities is certain to be military. This is clear from present trends: NASA was able to get funding for the Space Shuttle only by including military applications in the proposal, and a certain portion of Shuttle missions are reserved for secret military purposes; further, Reagan's "star wars" proposal shows that space is now seen as an important arena for weapons development. An extension of the military rivalries between nations into space is likely to make peace more precarious, and thus to lower significantly the overall level of human welfare.

Thus, the prima facie case holds, and is even strengthened. It is clear that a public policy of space colonization would not be morally justifiable. This conclusion, of course, depends on our argument that the pessimistic scenario is correct. But we can strengthen this conclusion even further by considering more carefully what the moral implications would be were the optimistic scenario correct. We must question our earlier assumption that a public policy of space colonization would not violate considerations of justice and fairness.

We argued above that under the optimistic scenario space colonization would be morally justifiable because the great amount of social benefit created for the generation living in space in 2200 would outweigh the present generation's loss in benefit required to initiate space colonization. But we asserted at the start of the paper that to be morally justifiable, policies must not only be defensible on social-benefit grounds, but must also satisfy concerns of justice and fairness. Under the optimistic scenario, the future generations in space would benefit at the expense of a loss in welfare to present generations. This raises the problem of justice between generations.

Obviously, the level of welfare of any generation is dependent on the degree to which previous generations have sacrificed themselves. But can it be morally correct to require, as the adoption of a public policy of rapid space colonization would, one generation to sacrifice greatly to improve the welfare of future generations to an even greater degree, even though this would be recommended on grounds of social benefit alone? Consider this analogy. Say that it were the case that the institution of slavery led to a much higher gross national product than would be the case were the economy to be run without slave labor. Surely slavery in this case could not be morally justified, even on the correct observation that a sacrifice by one group in society led to an even greater gain in benefit on the part of other social groups. Should not the argument be the same if we are speaking of different generations rather than different social groups within the present generation? A public policy of rapid space colonization would take needed resources from the present generation and give them to future generations. Such a resource distribution is as unjust as one which allowed an affluent leisure class to live off the labor of the poor.

So even were the optimistic scenario correct, the sacrifice in benefit that space colonization would require of the present generation is sufficient to show the policy to be morally unjustifiable. The case against the moral justifiability of space colonization holds whether it is the optimistic scenario or the pessimistic scenario that is correct. A public policy of devoting the present resources necessary for housing a great number of people in outer space is a policy of _not_ devoting such massive resources to other, more pressing problems confronting this generation of humanity. Adequate levels of food, shelter, and medical care are basic human needs that much of the world (and the nation) still lacks. Not every technological possibility is worth actualizing.

A final reminder of the analogy to medical technology may bring out the concern with justice and fairness more clearly. As we have seen, the use of public funds to develop very expensive life-saving devices that would benefit only a few people is as much a moral problem as it is an economic one. These devices _could_ be developed, but at the cost, given the limited resources available for health care, of failing to provide a large number of people with more basic medical necessities. The economic problem becomes a problem of distributive justice: since resources are finite, they ought to be allocated in ways that are fair.

CONCLUSION

Given the resources that would be needed for space colonization, a case must be made for the fairness of the sacrifices this would involve. We have argued this case cannot be made. The idea of space colonization has its uses as a device to help students to understand some important issues central to a liberal-arts curriculum, but one of these uses leads to the conclusion, if our argument is correct, that space colonization should not be adopted as public policy.

REFERENCES AND FOOTNOTES

1. Allan Russell, "Human Societies in Interplanetary Space: Toward a Fructification of the Utopian Tradition," Technological Forecasting and Social Change, 12, 353-364, (1978).

MOON COLONIZATION AND THE IMAGINATION: A PSYCHOLOGICAL INTERPRETATION OF ROBERT A. HEINLEIN'S "REQUIEM"

Barbara Eckman
Department of Religious Studies
University of Pennsylvania, Philadelphia, PA 19104

ABSTRACT

Heinlein's "Requiem" depicts Delos D. Harriman's longing for the Moon. Reaching his final resting place on the Moon is, to Harriman, coming home. Why does Harriman long for the Moon as one longs for home? This question is addressed using the interpretive method of James Hillman's archetypal psychology. To Hillman the Moon metaphorically represents the imagination. If Moon is imagination, then Harriman's desire to walk the surface of the Moon amounts to a desire to be "grounded" in the imagination; and his desire to view the Earth as a lunar "satellite" amounts to a desire to overcome the alienation between the imagination and the "natural [earthly] perspective," which recognizes as real only the physical and the noetic, neglecting the imaginal. This understanding of the longing for a home on the Moon suggests some potential benefits for actual Moon colonization.

INTRODUCTION

As its subtitle suggests, this paper presents a psychological interpretation of a science-fiction short story which deals with the topic of Moon colonization: specifically, the longing to set foot on the lunar surface. The paper also serves as an example of how the topic of space colonization might be included in an undergraduate liberal-arts course. For example, in the course on archetypal psychology (the vanguard of the psychological school which traces its origins to the work of C.G. Jung) that I am currently developing, space colonization might be introduced as an example of a powerful idea whose emotional impact may be elucidated by means of archetypal analysis. In this paper, which might be delivered as a lecture in such a course, my aim is to provide at least a partial answer to the question, "Why are stories about space colonization so moving and powerful?"

0094-243X/86/1480105-9$3.00 Copyright 1986 American Institute of Physics

A second introductory remark concerns the principles of interpretation used in this study: specifically, whether the author's
intention strictly determines the meaning of a text. While the issue
of the author's intention fully warrants the in-depth consideration
it has received in the writings of a wide variety of philosophers and
literary critics, a brief note must suffice here. I do not claim to
be stating what Heinlein, the author of the story, intended by his
use of the theme of Moon colonization. Rather, I am interested in
investigating the power that the text of the story has on contemporary readers.

Finally, a synopsis of the plot of "Requiem" (1939) may prove
helpful here. Delos D. Harriman is actually "introduced" in a later
(1949) Heinlein story, "The Man Who Sold the Moon."[1] In that story
Harriman, following a lifelong dream, singlehandedly convinces Terran
entrepreneurs and governments to undertake the work of Moon colonization. But Harriman, as financial genius of the Moon colonization
effort, is not permitted by his partners and stockholders to
undertake the hazardous Earth-Moon journey himself, because he is
deemed too valuable an asset to be risked in this way.

"Requiem" takes place many years later, long after Luna City,
the first Moon colony, has been established. Harriman's dream of
Moon colonization is now a reality. His financial partners no longer
bar him from the Earth-Moon flight. But now, an old man with a very
weak heart, Harriman cannot obtain medical clearance to make the
journey. Desperate to reach the Moon in the little time remaining to
him, Harriman engages a carnival rocket pilot and his engineer to
secretly and illegally build a private ship capable of the Earth-Moon
run. The large-scale liquidation of financial holdings required to
raise capital to construct the ship causes Harriman's heirs to seek
legal certification of Harriman's mental incompetence. When it
becomes apparent that their petition will be granted, Harriman flees
to the secret desert site of the ship's construction. The ship,
christened the Lunatic, takes off from Earth just as the police
arrive with a warrant for Harriman's arrest for contempt of court.

Thus Harriman accomplishes the voyage of a lifetime. But the
massive forces of acceleration involved in takeoff and landing take
their toll on his weak heart. He is barely breathing when the crew
remove him from the ship. But he survives long enough to feel the
lunar soil in his hands, and to view the Earth at last quarter over
the western horizon.

> At long last there was peace in his heart. . . . He was
> where he had longed to be -- he had followed his need.[2]

Thus content to rest on the surface of the Moon, Harriman dies.
His crew make his grave in the ashen lunar soil, leaving behind only
a scrawled epitaph -- a testimony to Harriman's great longing and to
the final fulfillment of his desire.

THE MOON AS HOME

Under the wide and starry sky
Dig the grave and let me lie
Glad did I live and gladly die
And I laid me down with a will!
This be the verse you grave for me:
'Here he lies where he longed to be,
Home is the sailor, home from the sea,
And the hunter home from the hill.'[3]

These words comprise the epitaph that Robert Louis Stevenson wrote for his own grave. Though a Scotsman by birth, Stevenson wished to be laid to rest on a high hill in Samoa. Similarly, the Moon was Harriman's final resting place and the home of his heart. A deep longing like Harriman's longing for the Moon presupposes a prior alienation. What did earthbound Harriman lack? How did Moon colonization answer his need? How might it heal our alienation?

In this paper I will investigate some of the metaphorical[4] content of Moon colonization, using the psychological theory of James Hillman, a Jungian analyst and the leading exponent of archetypal psychology.[5] In speaking of metaphor I do not mean to imply that current interest in Moon colonization is "merely" metaphorical -- far from it! But if history is a reliable teacher, pioneering and colonization movements are likely to possess great metaphorical or even religious meaning. In Hillman's terms, such movements are "archetypal": they possess "value, [an] all-encompassing sense of importance."[6] Consequently, it behooves those who are contemplating such pioneering projects as space colonization to be aware of this archetypal content. Such awareness may prove beneficial in a variety of ways, from simply enhancing human self-understanding to producing effective public relations campaigns aimed at stimulating popular interest in colonization. But more on this later.

The question of how the Moon can be the home for which the earthbound long may be addressed from a number of perspectives, even within the limited framework of archetypal psychology. If one focuses on the crescent Moon, which Hillman calls "the tiny virgin girl at the beginning,"[7] then Moon colonization might signify a longing for a return to lost innocence. On the other hand, if the Moon is identified with the Hebrew mythical figure Lilith, Adam's banished first wife turned baby-eater, then longing for a home on the Moon might involve a desire to recover and redeem the repressed feminine qualities or values which have been denigrated in patriarchal culture. For the purposes of this paper I prefer to focus on the full Moon, in which the various features of the lunar landscape are discernable; for this is the face of the Moon that D.D. Harriman favors in Heinlein's story:

He had sat out on the veranda all night long, watching the full Moon move across the sky. . . . Tonight he'd be alone with his old friend. He searched her face. Where

was Mare Crisium? Funny, he couldn't make it out. He
used to be able to see it plainly when he was a
boy. . . . But he didn't need to see, he knew where they
all were; Crisium, Mare Fecunditatis, Mare Tranquilitatis
-- that one had a satisfying roll! -- the Apennines, the
Carpathians, old Tycho with its mysterious rays.[8]

According to Hillman, this full Moon is metaphorically
identified with the imagination. In speaking of the metaphorical
value of silver, which in alchemy is "the metal of the moon,"[9]
Hillman says:

Imagine this metal then as a nontangible whitened air, a
silvering white body, ethereal like the orb of the full
moon floating, suspended in dark azure receptivity, a
hard, cool and bright mind at its full.[10] Silver
presents the moon's full brightness, its completion or
elevation. It is the metal of a Great Light, to use the
traditional way of speaking, . . . the anima [soul]
realized as an imagination so solid, a soul so bodied,
that the reflections of its images whiten the earth of
mundane consciousness [as reflected moonlight whitens the
literal Earth].[11]

By "imagination" Hillman means the quintessential activity of
the human psyche or soul. This "soul" is not to be confused with the
Greco-Roman or Judeo-Christian "spirit," which is defined primarily
by its opposition to or distinction from body. Rather, soul is a
"middle term" between spirit and body, in opposition with neither and
relating freely with both. Hillman defines imagination and psyche
circularly: imagination is the quintessential activity of psyche,
and psyche is first and foremost the seat of the imagination and the
wellspring of archetypal images and fantasy.[12]

THE MOON AS HEALER OF EARTHBOUND ALIENATION

To understand more specifically how the imagination can heal
one's earthbound, alienated condition, let's look at Harriman's
stated reasons for wanting to reach the Moon:

It's the one thing I've really wanted to do all my life
-- ever since I was a young boy. . . . When I was a kid
practically nobody believed that men would ever reach the
Moon. . . . But I believed. . . . I believed that we
could do it -- that we would do it. I set my heart on
being one of the men to walk the surface of the Moon, to
see her other side, and to look back on the face of the
Earth, hanging in the sky. . . . I've lived longer than

> I should, but I would not let myself die. . . . I will
> not! -- until I have set foot on the Moon.[13]

As this quotation shows, Harriman wants three things:

- to walk on the Moon;
- to see the so-called dark side of the Moon;
- and to look back on the Earth from the vantage point of the Moon.

Using Hillman's theory and method, I will dwell for a moment on the metaphorical or, as Hillman says, the "imaginal" content of these three desires.[14]

To Walk on the Moon

The wish to walk the Moon's surface, to set foot on the Moon, connotes a desire to be "down to earth" in the ways of the Moon — to be solidly grounded in the ways of the imagination. Says Hillman

> The grounding of mind in whitened earth [= Moon rocks,
> the lunar imaginal perspective] is what I have written
> about at other times as the poetic basis of mind. . . a
> reflection of images, an ongoing process of poesis, the
> spontaneous generation of formed fantasies. . . . For
> these ores of the moon, these mythopoeic materials, are
> the primordial seeds out of which soul life comes.
> Platonic psychology says that souls originate on the
> moon. . .[15]

To be grounded in the imagination means that one recognizes that human fantasies and poetic images are as real as physical and noetic or intellectual reality. To be so grounded is to be competent in, to know intimately and responsibly, the ways of the soul or psyche: not to be flighty, spacey, or "airheaded,"[16] nor credulously to "swallow whole" whatever a fantasy might suggest, so as to abdicate conscious responsibility; but rather to know the value of images and fantasies as intimately and in as disciplined a manner as poets do, and to bring the insights of the imaginal perspective to the experience and shaping of the everyday world of love and work.[17]

To See the Moon's "Dark" Side

Harriman's second desire, the wish to see the Moon's dark side, is a desire to witness "the aspect of Luna that turns its back on the earth and does not participate in daylight concerns,"[18] the place whence souls originate. Harriman wants contact with the shadowy place, oblivious to mundane concerns, whence imagination comes, where psyche arises.[19]

To View the Earth from the Moon

Finally, Harriman wishes to look back on the Earth from the perspective of the Moon. From this vantage point, the Earth appears "hanging in the sky" as "a green-blue giant moon,"[20] and the soil of the Moon provides the "grounding" underfoot. Earth has become Moon, and Moon has become Earth. If metaphorically the Earth represents "the mundane world of the natural perspective"[21] and the Moon represents the "unearthly" process of poesis or reflection of images, then imagination has become ground on which to stand, and everyday concerns have become more "airy," less deadly serious and literal, less dominating of consciousness, and more willing occasionally to play a satellite role to imaginative concerns. In Jungian terms a conjunction has taken place between Earth and Moon, between the "natural" perspective and the "unearthly" one.[22] As Hillman puts it,

> When the mind rests on imaginal firmament, then thinking and imagining no longer divide against each other. . . . Knowledge comes from and feeds the soul. . . . Here is knowledge not opposed to soul, different from feeling or life, [narrowly] academic, scholarly, sheerly intellectual or merely explanatory (erklärendes), but knowledge as a necessity demanded by the silvered mind by means of which the soul can understand itself.[23]

THE CONJUNCTION OF MOON AND EARTH

On the deepest level what Harriman longs for, and what may be the archetypal basis of current interest in Moon colonization, is the conjunction of Moon and Earth, that is, the state in which knowledge and imagination, everyday life and the "underworld" of poetic shadows -- even technology and the liberal arts -- are no longer opposed to one another. The human self is no longer alienated from the deepest springs of its own life, but has "come home" to itself.[24]

For both Harriman and Hillman, then, living on the Moon signifies healing a condition of alienation from the life of the imagination. The "moral" of the story as I have interpreted it is to bring knowledge and imagination, realism and fantasy, back together, both on Earth and in space colonies: by drafting educational curricula that can contain the soul's imaginative and poetic needs, as well as the mind's noetic ones; by constructing public transportation systems that also transport the soul through their aesthetic design -- and in countless other ways. I see this conjunction of Moon and Earth already occurring among scientists who do not shun the poetic imagination, and humanists who do not fear scientific investigation and technological advancement. I hope conferences like this one will encourage and expand consideration of "real world" ways of how these two real worlds, the physical-noetic and the imaginative, might come home to one another -- and already have -- whether in Philadelphia or New York or in Luna City.

FOOTNOTES AND REFERENCES

1. In Robert A. Heinlein, The Past Through Tomorrow, (Putnam's Sons, New York, 1967).
2. Heinlein, p. 261.
3. Robert Louis Stevenson, "Requiem." Quoted in Heinlein, p. 245.
4. Those seeking a precise definition of metaphor in Hillman's writings will be disappointed. Nonetheless, an illuminating discussion of the metaphorical as distinct from the literal may be found in James Hillman, Re-Visioning Psychology, (Harper & Row, New York, 1975), pp. 149-154. In Hillman's usage, the metaphorical differs from the literal precisely in its "permanent ambiguity"; the literal, on the other hand, narrows "the multiple ambiguity of meanings into one definition." In avoiding a precise definition of metaphor, then, Hillman's writing remains metaphorical.
5. I will be relying primarily on the following essays: "Silver and the White Earth (Part One)," Spring: An Annual of Archetypal Psychology and Jungian Thought 1980, 21-48; and "Silver and the White Earth (Part Two)," Spring, 1981, 21-66. Hereafter I will refer to these essays as Silver I and Silver II, respectively.
6. James Hillman, "An Inquiry into Image," Spring, 1977, 75; see also 82-83.
7. Silver I, 25.
8. Heinlein, p. 250. The use of the feminine pronoun to refer to the Moon in this passage suggests that a more traditional Jungian interpretation of the Moon as Harriman's anima or unconscious feminine "soul-image" would also be fruitful, particularly since Harriman fled to his veranda because of a fight with his singularly unimaginative wife, who could not understand his need to reach the Moon. Such an interpretation would, however, not contradict the present interpretation of the Moon as imagination, but rather support and complement it.
9. Silver I, 22.
10. Silver I, 24 (my emphasis).
11. Silver I, 25.
12. This creative imagination as the quintessential activity of psyche is the foundation for Hillman's theory of "the poetic

basis of mind." See Silver I, 46 and note; also quotation of reference 15 below.

13. Heinlein, pp. 249-250. Excluding "that we <u>would</u> do it," all emphasis is mine.

14. The word "imaginal" was coined by the Islamic scholar Henri Corbin; see his "Prefatory Letter" in <u>The New Polytheism</u> by David L. Miller, (Spring, Dallas, 1981), p. 4:

> I have for many years endeavored to restore both logically and gnoseologically a mediating and inter-mediary world which I call <u>mundus imaginalis</u> . . . . This is an <u>imaginal</u> world not to be confused with the <u>imaginary</u>.

On Hillman's equation of "imaginal," "metaphorical," and "psychological," see, for example, <u>Re-Visioning Psychology</u>, pp. 149-154.

15. Silver II, 25-26.

16. <u>Cf</u>. Silver II, p. 24:

> Our heads are always reaching up and out to the celestial earth [the Moon]. And the problem of head trips is not that they are trips or that they are heady, but that they are not grounded. To ground these flights of fancy and ideational excursions, psychology sends the head down again to the material earth, insisting it bow down to the dark madonna of tangible concrete existence. Psychology fails to grasp that spacey head-trips are in search of another ground way out, an attempt to reach the u-topic, placeless, farfetched terra alba [whitened Earth] of anima inspiration.

17. That is, to perceive the "imaginal potential" already inherent in mundane experience and physical reality. See James Hillman, "Anima Mundi: The Return of the Soul to the World," <u>Spring 1982</u>, 89. "Love and work" is the traditional Freudian recipe for a healthy life.

18. Silver I, 39.

19. <u>Cf</u>. Silver I, 46:

> The psyche comes to each moment of the world from the moon -- not just once at birth in a mythology of creation, but at birth each day, right now.

20. Heinlein, p. 261.

21. Silver II, 26.

22. For Hillman the "natural perspective" includes the noetic as well as the physical: both the intellectual and the "down-to-earth," both mind and body. It comes naturally to us to see the human person as a "mind-body problem"; but the middle term, soul or imagination, is given no place in our culture's intuitive self-understanding.

23. Silver II, 51.
24. <u>Cf</u>. Silver I, 38:

> The hardening of fantasy into obdurate psychic fact requires cooling. . . . The coldness here is not bitter, but relieving, as if the psyche comes home to itself, losing some of its airy flightiness, its attraction for the flame, its succulent attachments, becoming more firm and solid, achieving its kind of earth by becoming cold.

HOMESTEADING AND THE CREATION OF PROPERTY RIGHTS IN OUTER SPACE

Joel D. Scheraga
Department of Economics, Princeton University, Princeton, NJ 08544
Department of Economics, Rutgers University, New Brunswick, NJ 08903

ABSTRACT

The rapid proliferation of space travel has made the colonization of space and the exploitation of its scarce resources imminent.
The development of scarce resources in space has raised serious
legal, political and economic questions that must soon be resolved if
countries are to coexist peacefully in outer space.   This paper
examines the problem of defining property rights in outer space.
Private property does not yet exist in space.   This failure to
establish property rights is critical because the scarcity of
resources in the absence of ownership will inevitably lead to
problems of congestion, inefficiency, and socially undesirable
outcomes.   A program of homesteading, combined with a competitive
bidding system, is presented as an efficient and equitable means of
distributing resources in space.   The international debate over alternative schemes for allocating resources in space provides an opportunity for students to see the interaction of diverse but related
disciplines in the liberal arts.

INTRODUCTION

The development of the U.S. Space Shuttle and alternative space
transportation systems has made the colonization of space and the exploitation of scarce resources in outer space imminent.   Plans
already exist for the construction of a U.S. space station, the establishment of permanent bases on the Moon, a manned expedition to
Mars, and the mining of asteroids.   But the rapid exploration and exploitation of space has raised serious legal and economic questions
that must soon be resolved if countries are to coexist peacefully in
outer space.   Students in the undergraduate curriculum will find that
the international debate over these issues involves diverse arguments
from various liberal-arts fields, including economics, history,
political science, and sociology.

The purpose of this paper is to examine the problem of defining property rights in outer space. With the proliferation of space travel, many of the resources found in outer space are rapidly becoming scarce; that is, they are desirable but limited in quantity. The issue of scarcity is critical because it implies the need to allocate the resources among unlimited human wants and desires -- the fundamental problem of economics. It implies the need to establish property rights to the scarce resources.

Property rights have not yet been clearly defined in outer space. This failure is critical because it will inevitably lead to inefficient and inequitable distributions of space resources, international controversies and socially undesirable outcomes.

The international controversies that may develop due to the failure to define property rights in space are illustrated by the debate over the allocation of the scarce orbital slots for geosynchronous satellites. Satellites that are placed in geosynchronous orbit remain in a stationary position over a particular point on the Earth. These orbital positions are desirable because they permit a continuous flow of information to be transmitted between any two points on Earth within the satellite's range. The number of available slots, however, is limited to prevent interference between the transmissions of adjacent satellites. The United Nations' International Telecommunications Union currently attempts to allocate the slots between countries. The price of an orbital slot, however, is zero. Countries like the United States and the Soviet Union, which have large endowments of technology and wealth, are therefore rapidly appropriating most of the orbital slots. But in 1976 several equatorial countries signed the Bogota Declaration, staking a claim on the orbital slots directly over their countries. They asserted that the slots are an extension of their territorial "air space." The Bogota Declaration is not recognized by the world community, but the issue of who owns the orbital slots has yet to be resolved.

The international community is faced with a similar legal dilemma with regard to the colonization of the Moon. The Moon offers abundant oxygen that can be mined from rocks on its surface, materials that can be used for constructing facilities in space, and a base for astronomical observations in the absence of any intervening atmosphere. But international laws that will govern the establishment of lunar bases and the allocation of sites on the Moon have yet to be worked out. Fundamental differences still exist in the international community over the sovereignty of property on the Moon and in outer space. Unless these issues are resolved, future colonization and development of the Moon may be stunted.

The problem is one of ownership. Private property does not yet exist in outer space, and this failure to establish property rights is leading to inefficient and inequitable distributions of resources.

CONFLICTING SOCIAL OBJECTIVES

Conflicting social objectives in the international community have exacerbated the debate over resolutions to these problems. In fact, the international community has taken steps to prevent the establishment of private property in space.[1] In 1967 the United Nations drafted a "Treaty on Principles Governing Activities of States in the Exploration of and Use of Outer Space, Including the Moon and Other Celestial Bodies" (also known as the Outer Space Treaty). This treaty, which was ratified by 107 of the 159 member nations, provides the current framework for space law. It asserts that outer space is not subject to appropriation by any individual country.

At issue are two conflicting philosophies which are of interest to students of political science and space law: the "common heritage of humankind" versus the "province of all mankind."[2] The "province of all mankind" argument, upon which the Outer Space Treaty is based, asserts that all nations have equal rights and equal access to the resources of space. Although no country can claim ownership of a section of space, any country can indiscriminately exploit its resources. The conflicting philosophy extends the principle of non-appropriation, declaring all celestial bodies and their natural resources, as well as the trajectories around and beyond these bodies, the "common heritage of humankind." A possible implication of this philosophy (which is also embodied in the United Nations' Law of the Sea Treaty) is that all nations have a claim on a share of the profits from the development of a resource in outer space, regardless of who was the developer of the resource.

Significant disagreement exists over which principle should provide the framework for space law. Although the Outer Space Treaty embodies the "province of all mankind" argument, efforts have been made to enact space treaties that reflect the common heritage principle. One such treaty was the "Agreement Governing the Activities of States on the Moon and Other Celestial Bodies" (also known as the Moon Treaty). Although this treaty was adopted by the U.N. General Assembly in 1979, only five member nations have ratified it.

The eventual outcome of this debate is still unclear.

CONGESTION IN SPACE: WHY THE FAILURE TO ESTABLISH PROPERTY RIGHTS IS CRITICAL

As space is colonized, it is inevitable that problems of congestion will occur if property rights are not established. In the absence of property rights, the price of exploiting a scarce resource (such as desirable locations for settlements on the Moon and orbital slots for geosynchronous satellites) is zero. Students of economics will recognize that the opportunity cost to any nation of colonizing a particular location is lower than if property rights were assigned, so that scarce locations will be overused and overcolonized.[3]

The failure to define property rights leads to a divergence between the private costs faced by an individual nation and the social costs to all nations in the world community. In the absence of private property, a country colonizing areas in space will not fully take into account the "external costs" that it imposes on all other nations that may also want to exploit these locations. The colonizing country has no incentive to consider the social cost of exploiting another scarce location. It will consider only its own private cost of the colonization project.

To understand this point better, consider the expected future colonization of the Moon. The far side of the Moon offers scientists an ideal location for the placement of astronomical telescopes that would probe the universe. The absence of a turbulent and filtering atmosphere permits telescopes to scan the ultraviolet and infrared regions of the spectrum that are unobservable on the Earth. Radio telescopes on the far side are protected from the abundance of radio noise emanating from the Earth.

Now suppose a country has decided to place a large nuclear-waste disposal site on the far side of the Moon, rather than in some alternative location in space. When the disposal site is constructed, it imposes a nonpecuniary externality (or external effect) on all other countries that are interested in building and occupying lunar bases in this region. By building the disposal site, the country adds to congestion in the area, and appropriates a location that could have alternatively been used for scientific purposes.

The external effect on any one country is small, but the total effect summed over all countries is large. The country building the waste-disposal site, however, does not consider this total effect. It does not consider the social cost of occupying the scarce lunar location. It considers only the average cost of constructing the lunar garbage dump.

> **Proposition 1:** Each individual country that is colonizing outer space, acting in its own self-interest, will not make socially correct decisions when the scarce locations being colonized are not owned by anyone.

The resolution to the problem is straightforward. Adam Smith's generalization, as applied to scarcity problems in outer space, asserts that if the rights to scarce resources in space are assigned unambiguously to a particular country, and if free exchange of the rights is permitted, then these resources will be used efficiently.

Proposition 2:   Particular resources in outer space are
scarce.   Scarcity in the absence of ownership leads to
congestion and inefficiency.   The establishment of
property rights will lead to socially correct and
efficient uses.[4]

HOMESTEADING AS AN EFFICIENT AND EQUITABLE MEANS OF DISTRIBUTION

Numerous alternative methods exist for the creation of property
rights and allocation of scarce resources.   One possible method is
homesteading.

The concept of homesteading was originally established by the
Congress of the United States in 1862 when it enacted the Homestead
Act.   Free land in the West was offered to any U.S. citizen or
intended citizen, 21 years of age or over, who would settle the land.
Homesteaders were required to develop the land for five years before
receiving title to it.   A commutation clause enabled the homesteader
to acquire the land through cash purchase after six months of
residence.

Homesteading is a tested means for defining property rights in
regions where they did not previously exist.   The Homestead Act established property rights in previously ownerless public land areas,
abetted the development of new communities, and lessened the chances
of social and civil disorder by giving ownership to the occupants of
the land.   It rewarded the pioneers and developers of previously
unexplored areas by providing them with first rights to the land.   In
this sense, the Homestead Act was equitable.

An equitable and efficient framework for establishing property
rights in outer space can be developed by combining a system of
homesteading with a competitive bidding system.   Desirable locations
in space could initially be distributed through homesteading.   First
rights would go to those countries that pioneer and develop
previously unexplored areas.   Once the area is worked for a certain
number of years, the "homesteading" country would gain the property
rights to it.   Following this initial distribution, the property
rights could then be sold and traded in a competitive market.
Trading of the rights in a free market would guarantee that the
resources would be allocated efficiently because they would
ultimately end up in the hands of those countries that value them the
most.

It does not matter from an efficiency standpoint who initially
owns the scarce locations in space.   An outcome is efficient if one
country is unable to make itself better off without making another
country worse off.   (This idea of efficiency is central to economic
thinking about society's happiness.)   In a free-market economy in
which deals can be made, the rights to the scarce locations will end
up in the hands of those countries that value them the most.   But the

initial distribution of property rights does affect the <u>equitability</u> of the final outcome of the exchange process. To speak of equity is to speak of "fairness." An evaluation of the equitability of a market outcome must be based on intersocietal comparisons of utility and social-value judgments on the part of some bureaucratic authority. It requires normative judgments by sociologists, political scientists, and professionals of other disciplines of the liberal arts.[5]

Homesteading provides one possible initial allocation scheme that is based on one particular notion of equity -- the awarding of first rights to the pioneers of previously unexplored areas, regardless of their ability to outbid other countries for these rights. These pioneers will not necessarily be those countries endowed with the most wealth and technology. Even the most wealthy countries like the United States and Soviet Union are not able to undertake all proposed space projects. This will leave many projects for lesser developed countries to pursue as they develop greater financial and technical capabilities. China, Brazil, India, France, and West Germany, for example, are already developing the technical expertise to pioneer and exploit previously unexplored regions of space.

> **Proposition** 3: Clearly defined property rights provide sufficient incentives to preserve scarce resources. Homesteading provides an **equitable** method for establishing ownership in previously ownerless areas of space. Subsequent trading of these rights through a competitive bidding system also guarantees that the resources will be allocated <u>efficiently</u>.

CONCLUSIONS

If countries are to coexist peacefully in space, then the creation of property rights is inevitable. The fundamental differences that exist in the international community over sovereignty in outer space must be resolved before space colonization can proceed in any systematic manner.

Resolution of these issues will require students from different disciplines to exchange viewpoints and alternative approaches and methods. Students of economics have to study alternative schemes for efficiently allocating the scarce resources of outer space. Sociologists and political scientists must address issues of equity in the distribution of these resources. Issues of space law as it applies to the exploitation of resources and the development of new technologies must be debated. Historians can provide the insights gained from similar problems that previously have been encountered.

One option is to recognize that the market system can apply to scarce resources in outer space as well as to resources on Earth and

that a system of homesteading offers an equitable method for
initially allocating scarce resources in space. After the initial
distribution, competitive bidding for the property rights would
ensure an efficient outcome. Of course other methods of distribution
and allocation exist. The analysis of these will provide a rich opportunity for students of all disciplines in the liberal-arts
curriculum to understand alternative (and sometimes conflicting)
viewpoints, to apply the concepts and methods of their disciplines to
problems critical to space colonization, and to formulate original
public-policy prescriptions. In this way, they may better understand
and more clearly define the rights and responsibilities of all
nations.

REFERENCES AND FOOTNOTES

1. Daniel Deudney, <u>Space: The High Frontier in Perspective</u>
    (Worldwatch Institute, Washington, D.C., 1982), p. 46.
2. Robert C. Cowen, "Shackling Lunar Developers," Technology Review
    <u>88</u>, 6, 77 (1985).
3. Joel D. Scheraga, "Curbing Pollution in Outer Space," Technology
    Review <u>89</u>, 8-9 (1986).
4. Ronald H. Coase, "The Problem of Social Cost," J. of Law and
    Economics <u>1959</u>, 1-40.
5. A system of homesteading provides only one possible resolution to
    the equity issue. Alternative methods of allocation do exist.
    (For example, see: Joel D. Scheraga, "The Establishment of
    Property Rights in Outer Space," The Cato Journal <u>1986</u>.) These
    include quotas and licensing, taxation, allocation through
    decisions of the World Court, and alternative bidding systems.
    For example, India has proposed a licensing and bidding system
    for allocating the scarce orbital slots for geosynchronous
    satellites. Each country would initially be awarded a certain
    number of slots. The rights to these slots could then be bought
    and sold in a free-market economy.

CHAPTER 3

TECHNOLOGY AND SPACE COLONIZATION

This chapter consists of two papers and a panel discussion. Endorsers of the efficacy of technology are closely questioned, and a deep division on the issue becomes explicit.

SPACE BEYOND VAUDEVILLE

Arthur Kantrowitz
Dartmouth College, Hanover, NH 03755

ABSTRACT

The U.S. (wo)manned space program has been designed for broad popular appeal and perhaps this is a political necessity. This paper explores the possiblities for a much larger scale human expansion into space. Some opportunities for dramatic cost reductions coming first from wider competition and second from new technological possiblities are outlined. Finally, we recall the possible benefits of this enlargement of humanity's scope which were foreseen when "technological optimism" was in vogue.

INTRODUCTION

Everyone knows that space is terribly expensive. Remember the media calculation that a glass of water on the Moon cost $200,000. That powerful point helped to kill the Apollo program as soon as it ceased to draw a massive TV audience. What can justify a large space program after the TV audience switches off Shuttle landings and handsome astronauts tumbling at zero g? Is there space beyond vaudeville? I will attempt to answer this question in the dull traditional manner -- comparing costs and benefits.

COSTS

I.

The key cost which determines the magnitude of realistic possibilities in space is the cost of transportation to low Earth orbit (LEO). It is very easy to calculate the energy needed to put a satellite into LEO. Centrifugal forces balance gravity at a velocity of about 8 km/s. In economic energy units, say kilowatt-hours (kWh), the energy of one pound moving at this speed (plus the smaller energy required to raise it above the atmosphere) comes to about 4.5 kWh or about 30 cents worth of electricity. A Shuttle launch costs about $200 million now, and as of now the largest cargo to LEO is about 55,000 lbs. Thus current cost for orbital energy on the Shuttle is about $3600/lb. The purpose of this calculation is to exhibit that in fundamental energy terms there is room for improvement by a very large factor.

In view of this large factor how have we been doing? The answer is to my mind the most discouraging aspect of our adventure into space. In the first quarter century the costs of transportation to LEO have not declined! This is in such stark contrast with almost any young industry that we should look at it very hard. Let's begin by surveying briefly the opportunities for bringing the cost of transport to LEO closer to the energy cost.[1]

II.

The business of transportation to Earth orbit has finally become competitive. Several countries, e.g., France, China and the U.S.S.R., have entered the international launch services marketplace. The development of launch vehicles is underway in Japan, Brazil, India and elsewhere. The emerging competition should reduce the cost of transportation to LEO. But competition could play a much more important role in forcing NASA and the other government monopolies which control space to look harder for technological opportunities to improve dramatically costs to LEO.

Two possibilities employ chemical rockets designed to minimize cost. In view of the fact that existing launchers are designed to military specifications or to carry human beings, it is likely that considerable savings could be had for payloads where more tolerance for launch failures is acceptable. Near term examples are transporting bulk commodities to the space station or launching satellites manufactured in large quantity so that a launch failure would not be so expensive.

The first approach was known as the Big Dumb Booster. Here it was proposed to use advantages of scale to achieve cost savings. It was thought that certain fixed costs, e.g., guidance, tracking, etc., could be spread over larger payloads and that manufacturing tolerances might be increased at the larger scale, leading to lower costs per pound. This proposal was offered to NASA by the Chicago Bridge and Iron Co. at about the time that the Shuttle was being sold to the Congress. Being competitive with the party line it naturally received its just deserts -- oblivion.

The second chemical rocket approach uses a booster to be assembled as much as possible from hardware manufactured in large quantity for commercial uses. For example, by using a dozen or more modules sized to be within available tubing sizes, commercial tubing can be used for rocket casings. Designs for guidance, telemetry, and

other electronic systems can profit from today's extremely
economical, diverse, and reliable commercial solid-state devices.
Development costs can be minimized by using nearly identical modules
so that most debugging can be done on a small scale using a single
module.

The possibility of reducing development costs to within reach of
many venture capital organizations has resulted in the formation of a
number of small companies (some starting up in Silicon Valley) who
can see profits while reducing the cost and increasing the access to
LEO for communication and other commercial satellites.   I am
persuaded that some of these small firms will succeed and prosper in
those countries which encourage this activity by making launch,
tracking, and other facilities available.   At present there is some
indication that the U.S. is encouraging these firms and this is very
favorable for U.S. prospects for leadership in the large-scale use of
space.   I would anticipate that this development could reduce the
cost of transportation to LEO by a factor of two or more in the next
decade.

III.

There are a number of more radical approaches to reducing the
cost to LEO for large-scale space operations.   In several of these,
large electrical energy sources on the ground provide power for the
acceleration of vehicles to orbital velocity.

Pulsed lasers can easily heat targets to temperatures sufficient
to vaporize any material.   Indeed it is possible to transfer energies
large compared to chemical energies, to the evaporated material.   The
evaporated material produces a jet which propels the target and the
kinetic energy of the propulsive jet can be a large fraction of the
energy absorbed from the laser.   If the vapor is heated to very high
temperatures, orbital velocities can be achieved without the expendi-
ture of large amounts of propulsive material.   The laser which
supplies the energy remains on the ground.   To accelerate the target
vehicle over a long range, a very large laser is required (average
power of the order of a million kW).   Thus a large front-end cost is
required and this means a large commitment to a new technology.

The great advantage of laser propulsion is the dramatic
reduction of the liftoff weight which results from the high jet
speeds achievable.   Thus it is possible that liftoff weights for a
given payload could be reduced by a factor of ten or more with
comparable reductions in the cost of transportation to LEO.   If
President Reagan's Strategic Defense Initiative is pursued, very
large ground-based lasers may well be developed, and laser propulsion
would be much closer to realization.

Another approach to reducing the cost to orbit is to accelerate
vehicles electromagnetically.   In one such approach a variant of the
electromagnetically suspended train is used to accelerate a vehicle
to orbital speeds while the vehicle is still at ground level.   Since
aerodynamic drag at orbital speeds is very large (several hundred at-

mospheres stagnation pressure) it is necessary to shape the orbital vehicle to minimize drag -- to make it very long compared to its diameter -- and to carry a considerable thickness of ablating material to protect the vehicle from aerodynamic heating which could be several times that encountered during reentry from orbit. Nevertheless this proposal should be studied carefully since it could offer order-of-magnitude cost savings in the transportation of bulk commodity materials to LEO.

I will close this brief survey of opportunities for technological savings with a still more imaginative version[2] of electromagnetic acceleration to orbit. The Launch Loop will exhibit the variety of possibilities which need study. The Launch Loop stores momentum and energy in a very long (2000 km) loop of iron ribbon moving at somewhat higher than orbital speeds which will lift it off the ground. Tethered by kevlar cables it is allowed to ascend to an altitude of about 80 km which puts it just above the dense atmosphere. The loop is accelerated with ground based electric power. The forces needed to shape the loop are provided by electromagnets close to the ribbon, and stability is provided by high speed control of the electromagnetic forces. If this control can be achieved, then vehicles can be launched by electromagnetic coupling to the moving ribbon. This system, which would require a substantial front end cost, might in the long run be the cheapest way to the large-scale use of space, but much invention, research, and analysis is required before that possibility can be assessed.

BENEFITS

I.

I want to do my best to expand discussion of the benefits of the space adventure from the search for "niches" which constrains our thinking too much today. I propose to return to the attitudes which were common before 1957 when people who thought about space at all considered that it was to be an important part of the future of the human race.[3]

But why is the search for niches so important today? Clearly in a pessimistic atmosphere where harsh criticism confronts all new technology, NASA must sell the space program either as harmless vaudeville or on the hope for short-term returns on investment. However, it is important to realize that selling the space program so cheaply feeds the pessimism which is our most important disease. Space becomes another failure of technology to live up to its promises. Discussion of the utility of space in terms of the search for niches surrenders to an agenda forced on us by established pessimism.

II.

Henry Adams[4] was struck by the consequences of the Second Law of Thermodynamics for humanity's future. Quoting Kelvin and Tait to the effect

> . . . that all of nature's energies were slowly converting themselves into heat and vanishing in space, until at last nothing would be left except a dead ocean of energy at its lowest possible level . . . incapable of doing any work whatever . . .

While this can't be regarded as anything but a metaphor, unless you have the ability to see billions of years into the future, the metaphor has been abused repeatedly by the antitechnology movement[5] in its effort to exhibit the futility of technological progress. I propose to see if the same metaphor can be used to exhibit the promise of expansion into space for humanity's future.

First a small lesson in elementary thermodynamics. As long as you can keep expanding a system you can turn heat at one temperature into work (notice that this is not a cyclic process and therefore no violation of the Second Law). For example, the work which can be obtained by doubling the volume of a mole of gas maintained by heat addition at temperature T is $RT*\ln 2$ (R is the gas constant) regardless of the initial volume. Thus, if you are approaching "a dead ocean of energy," it is still possible thermodynamically to maintain life by creating a new way to expand the system.

III.

I propose to try to draw your attention to the thesis that expansion into space is an opportunity to revitalize our civilization. I do not want to suggest that we are left with a dead ocean of energy, but it is clear that our growth and our optimism are not what they were in an earlier stage in our development or indeed what they are in some developing countries today. Let's start instead from the grand concept of the Dyson sphere. The distinguished physicist Freeman Dyson discussed the possibility that we could build a sphere surrounding the sun with approximately the radius of the Earth's orbit. Such a sphere would intercept two billion times as much sunlight as is received by the Earth. He pointed out that it would take something like the mass of Jupiter to build such a sphere. With an ecology based on continuous sunlight at terrestrial intensities and the recycling of all materials, the sphere could support a population unimaginable by terrestrial standards. The Dyson sphere will serve to distinguish what I have to say from the agenda of established pessimism.

IV.

But the Biblical injunction to go forth and multiply could certainly have a larger meaning than simply expanding the population. I want specifically to call your attention to the magnificent opportunity that space offers for new beginnings. America is, of course, the classical example of a new beginning. More easily than in older societies it was possible here to implement the grand vistas of the enlightenment. There was not so strong an aristocracy to be defeated before the new ideas could be heard and implemented. Terrestrial opportunities for new beginnings still exist, but now somebody must be displaced, as in the tragedy of Israel. The replacement of an established structure of vested interests requires so vicious a struggle that utopias are degraded into gulags.

The location of utopias in space needs only the power of what the pessimistic establishment calls technological optimism. But what a creative power that was, and what a power it could become if challenged to create the technology needed for utopias in space! I will hazard the prediction that the costs of the 4.5 kWh/lb of energy needed to accelerate passengers and their needed goods into orbit will then approach the costs of 4.5 kWh of electrical energy; then we will have the means for an immigration into space comparable to that into America. Space will provide opportunities for repeated attempts to try for improved social organization, and by trial and the elimination of error (Popper's great phrase) we will ascend to societies limited only by our imagination.

But technological optimism has been convicted and sentenced to permanent oblivion, and if the pessimistic establishment discovers that we have even been thinking such thoughts, they will be very -- silent.

REFERENCES AND FOOTNOTES

1. After the Challenger tragedy our decision to proceed with manned space should be expressed by a determination to develop the technology needed for safe, cost effective, large-scale transportation to Low Earth Orbit. The tragedy has called attention to the human price we are paying for a program which has not demonstrated that it will ever lead to the large-scale space operations in which people can live and work in the expanded world which space offers.
2. Lofstrom (private communication).
3. J.D. Bernal, The World, The Flesh, and the Devil, (Methuen, London, 1929).

4. Henry Adams, "A Letter to American Teachers of History," in The Degradation of the Democratic Dogma, (Macmillan, New York, 1919; reprinted by Peter Smith, New York, 1949), p. 145.
5. Cf. Jeremy Rifkin, Entropy: A New World View, (Viking-Penguin, New York, 1980).

RESOURCES AND RECOLLECTIONS OF SPACE COLONIZATION

Thomas A. Heppenheimer
Center for Space Science
11040 Blue Allium, Fountain Valley, CA 92708

ABSTRACT

The author presents a personal overview of and guide to the basic ideas of space colonies and space colonization. There is special emphasis on the developments that grew out of the NASA-sponsored summer studies of 1975, 1976, and 1977, instigated and led by G.K. O'Neill after he revived and stimulated interest in space colonization. The author describes how L5 was chosen as a site for a space colony; how the idea of powersats was linked to the idea of space colonies; how the L5 Society was founded; how the idea of using asteroidal materials became part of the picture; and how he wrote his books and developed certain of his ideas.

REVIVAL OF THE IDEA OF SPACE COLONIES

In the mid-1960s Dandridge Cole, along with D.W. Cox, published a book called Islands in Space which for the first time laid out the concepts of space colonization essentially as we understand them today. He drew on Arthur Clark's idea of the mass driver (if you want to call the electro-magnetic launcher that), and he drew on Darrell Romick's idea of large pressure shells. They got an artist to make some illustrations that considering the printing technology of the day are to be compared with those of Rick Guidice. Dan Cole really did a pretty good job with his book Islands in Space. (Chilton Press was the publisher and the book came out about 1964.)

Early in the 1970s at Princeton University, Jerry O'Neill reinvented the self-contained pressure shell and the mass driver concept. He wrote an article describing his ideas which a lot of journals turned down. Finally, in September 1974 Physics Today published it as "The Colonization of Space." He pointed out that it was possible to do all these wonderful things. What was more, O'Neill said one could have the first colony up there by 1988. Well, that was a possibility to consider.

There were a few other things that were happening at about that time as well. It was 1974, which might be considered the year 1 of the modern era of space colonization. There were actually, as in the French Revolution, only a handful of numbered years that meant anything.   In 1974 O'Neill went out to the West Coast where they have the Point Foundation, an affiliate of Stewart Brand, publisher of the "Whole Earth Catalogue." From the Point Foundation, Stewart Brand gave him $600 with which to put together a conference. O'Neill held that conference at Princeton and invited some of his friends. Gerald Feinberg, the physicist from Columbia, came and so did Walter Sullivan from the New York Times.   Sullivan put a story about this conference on the front page of his august, establishment journal.

ORIGINS OF L5

A couple of interesting things came out of that meeting. For example, one of the other participants, this was 1974, was a fellow at Princeton named George Hazelrigg. Those of us who know him know he was very well connected, very well tied in with whatever was going on at Princeton that was interesting. He had his fingers in a lot of pies. He showed up at the conference, and O'Neill was holding forth and saying "Well, we can put the colony at a libration point where the gravity of the Earth and the Moon cancels out because they are co-equally balanced. And we can put the colony there between the Earth and the Moon and that's a nice location."

Hazelrigg, who had been working professionally in these areas as O'Neill had not, spoke up and said "You don't want to put it there, it's not stable. You would burn up the weight of the station in fuel just keeping it on station." O'Neill said, "Oh, what do we do about that?" And Hazelrigg said, "Well you know there are stable libration points; you could put it at L4 or L5."   That impromptu conversation, not highly technical analysis, gave rise to the legend that L4 or L5 are places to place a colony.

Somehow L4 got dropped out and L5 wound up taking primary place.

LINKING POWERSATS TO SPACE COLONIES

One other thing happened in 1974. Down in NASA in Washington, a fellow by the name of Jesco von Puttkamer, who had a fondness for innovative or advanced schemes, came up with an important extension of Jerry's ideas. To wit, that O'Neill had up to that point mainly been reinventing Dan Cole's wheel, no pun intended as far as shape is concerned. Now von Puttkamer invoked the energy crisis. (Remember that?) He said, "You know there's this idea that's been kicking around for a few years ever since Peter Glaser up at Arthur D. Little came up with it: the solar-power satellite. Why don't we consider that we could build solar-power satellites in a space colony? The colony would then house the workforce that builds those powersats out of non-terrestrial materials."

On the strength of these developments von Puttkamer proceeded to put a modest amount of NASA seed money into the two major ideas of 1975. These were a large, not small, conference to be held at Princeton in the spring of that year, and the NASA-sponsored Summer Study that is the reason we are here today.

The 1975 conference in Princeton, which I was pleased to attend, took place in May and brought together a variety of people, many of whom Jerry had already known. Give the man credit. As a professor at Princeton he had a wide array of contacts and brought forth a number of ideas many of which were highly germane. Some of the papers were reasonably well written. This conference in the fullness of time gave rise to a volume of proceedings. In fact, the first of several volumes of proceedings.

One of the groups O'Neill was talking with was the American Institute of Aeronautics and Astronautics, the leading aerospace organization in the United States. Within that organization was Jerry Grey, who became interested, and he arranged to publish the proceedings of both the 1974 and the 1975 conferences, which appeared, finally, in 1977. (Who says that publication times are short?) The paperbound cover of that volume shows the Chesley Bonestell rendering of a mass-driver propelled spaceship approaching an asteroid.[1] The contents of the volume describe what took place in May at Princeton.

THE 1975 NASA/ASEE SUMMER STUDY

Then in late June, many of us gathered in NASA's Ames Research Center for the ten-week summer study which Harry Jebens has just eloquently described to us. That study, of course, gave rise to the first NASA publication on space colonies,[2] NASA SP-413. (It might very well be out of print. A lot of the stuff I'm telling you about here is out of print, which means that you have to haunt your library or talk to people who have this stuff and then be prepared to make great use of a Xerox machine.)

FOUNDING OF THE L5 SOCIETY

One other thing happened during 1975. Among the participants both at Princeton and also as a visitor to Ames, was a lady named Carolyn Henson. A very activist type, her father was a well-known astronomer. In fact both her parents were astronomers, Aden and Marjorie Meinel. She was out to make a name for herself and to contribute to establishing space colonies. She said, "Let's start up an organization. Let's call it the L5 Society. Let's publish a newsletter." All of this over the fullness of time came to be. She and her husband Keith Henson founded the L5 Society. Its original newsletter, which came out at the end of 1975, was four pages long and was a single sheet folded in two and printed on four sides. That was the Volume 1, No. 1 of the L5 News.

TWO BOOKS ON SPACE COLONIES

This was at a time when these ideas were new and exciting and they were getting an ever-widening following. At this point we began to see moves made toward the writing of books on this subject. Dan Cole wasn't good enough and we needed new stuff. Jerry O'Neill proceeded to write a book called The High Frontier.[3] I have not read that book; I therefore cannot tell you of its merits. I can tell you that so far as I understand from others who have read it, it is a reasonable statement of Gerard O'Neill's views on the subject. And on that basis I commend it to your attention.

I will now tell you of my book, Colonies in Space.[4] First, understand my situation in the fall of 1975. I was a research fellow at Caltech whose fellowship was about to expire, meaning they were not renewing it. Some of my colleagues on the faculty, all of whom had full professorships, didn't want to have me stay for another year. And when my professor who had brought me in, tried to get me a job at JPL (the Jet Propulsion Laboratory), they decided they didn't want me either.

Soon I would be going down every two weeks to the unemployment office of the State of California to tell them how I was looking for work and to receive in exchange a check for $180. That stood to be my future for 1976. I wanted to do something about that.

In October I went off to Tucson to spend a few days visiting with the Hensons. On the way driving back I meditated on these issues. I thought 'You know, this is a subject that deserves to have a book written about it. I could write that book.'

You see during that summer I had arranged with the AIAA, the American Institute of Aeronautics and Astronautics, to write an article for their journal Astronautics and Aeronautics, reviewing what the Summer Study had come up with. My article appeared in the March 1976 issue of Astronautics and Aeronautics.[5] In the course of working on that article, which I had done during the month of September, I was astounded at how much material there was. I found myself writing paragraphs where I wanted to write pages, writing pages where I wanted to write chapters. There was a great deal of material. And so, on that day late in October I said to myself, 'I could write this book.'

When I left Caltech and started winding my way through the unemployment line, one of the things I decided to do was pursue that
book.   Through various vicissitudes I acquired a literary agent who
knew Ray Bradbury and who arranged for me along with my friend, the
artist Don Dixon, on whom I frequently rely for artwork, to give Ray
a presentation on space colonies.   I went to Ray's house in Beverly
Hills early in 1976, and the slides that I showed were very similar
to those Harry has just shown. Of course I gave a discussion of what
was going on, and Ray was enthralled.   In fact, he was enthusiastic,
and he showed it in a very characteristic way.   I told him as our
evening ended, 'Ray, for the first time I've met someone who is more
childlike than I am.'   Well, he agreed to write the introduction to
the book, and that was a big step forward.
     My agent then went to New York and by the end of that month,
January of 1976, he was telephoning me and saying that one of the
senior editors of the industry, Ian Ballantine, was very interested
and was talking about a contract for $20,000.   All of a sudden it
looked like some wonderful opportunities were opening up.   The
evening that my agent gave me that telephone call I went off to the
movies to see "Mahogany," with Diana Ross.   You may remember her
singing 'Do You Know Where You're Going To?' "Mahogany" was the story
of a young girl who comes out of nowhere to make herself quite significant.   That movie and that song strongly resonated with me that
evening.
     Well, that particular prospect fell through. We did not publish
with Ballantine.   In the end we published with Stackpole Books, a
small firm in Pennsylvania, which gave us an advance of not $20,000
but $2,500.   Nowadays I get more money for writing individual
articles let alone a complete book, but then I was just starting out,
and that was how my book ultimately reached print.
     The writing of that book was somewhat interesting. For example,
people seem to be most interested in the several chapters where I
describe the actual business of living within a colony.   There is a
chapter on agriculture in a colony. I took that agriculture chapter
almost word for word from the paper by Keith and Carolyn Henson that
they had presented in Princeton the previous year.   I figured they
were friends of mine; they wouldn't mind.   In those days the
copyright laws were a little more lenient than they are today. So if
you care to make the comparison of the chapter on agriculture in
colonies with the paper on agriculture in the 1975 proceedings, you
will be welcome to write an academic paper titled "Plagiarism as a
Major Force in the Career of T.A. Heppenheimer."

THE 1976 SUMMER STUDY

     In 1976 O'Neill set up another summer study, which again took
place at Ames. He brought in some fairly significant people, people
with good talent. That year he also brought in as an assistant Brian

O'Leary. O'Leary proceeded to introduce an important new theme into space colonization, the idea of resources, not just from the Moon, but from the asteroids. The idea had been floating around. It wouldn't knock your socks off if you had been working in that community in those days. Brian took the idea and, giving people proper credit, he developed and ramified it and put it together in a comprehensive way. He published an article about it in Science[6] and generally developed it as a theme for his own research. It was a good thing that Brian did that. Asteroid resources turned out to be a significant idea.

The '76 summer study had some people who knew electro-magnetics and some people who knew about the other pertinent technologies. All in all we wound up with some very significant contributions.

Initially I was not invited to the '76 summer study. Why? I suspect it was because I was writing a book, and O'Neill was writing a book too. I think that he didn't want me to see all this wonderful information that was coming up.

There was a problem however. They needed to solve the question of how one uses a mass driver operationally. They needed to know the technical characteristics of the trajectories associated with a payload's launch by the mass driver. This was a very serious question. Unless we could show that the mass driver could be used operationally and be effective, the whole thing would be entirely pointless.

O'Leary had the appropriate technical background; Ames had the world's greatest computer of those days, the ILLIAC IV. O'Leary, with Dave Kaplan, a young student from Michigan, attempted to solve that problem as I, working on the side down in Southern California, sent them notes and equations to help guide them.

Brian couldn't do it, or at least he didn't do it at the time. Time was getting short. The summer study of 1976 was five weeks gone, they had only one week to go, and they needed to solve this problem or else the whole question of trajectories would go down the tubes. They gave me a phone call, "Come on up and work for us as a consultant." I came up to work with them as a consultant. Brian talked with me. He showed me what they were doing, and I could see that they had the major elements of what they needed on hand. It was a matter of being able to integrate them and smooth the joints. This was something I could do very readily, very straightforwardly.

It wasn't so hard. As I said, it was only a matter of taking stuff that those other fellows had already ginned up and rearranging parts of it and reformulating other parts. By Sunday night (I had come up on Friday) we were ready to run on the ILLIAC; and then they shut the ILLIAC down for the night.

We were back Monday morning. We ran our first run and got the data out. I examined the printout, particularly looking at what we call the sensitivity of the data. We wanted that sensitivity to be small. I had reasons from theory to believe that it should be zero. And if we could get zero sensitivity then we would have everything we could ever want.

I reduced the first data point, and the sensitivity was very bad. I reduced the second data point, associated with a geographical

location very close to that of the first. The sensitivity was considerably worse. I went on to the third data point, again moving a
little farther along the surface of the Moon to a nearby point
adjacent in longitude. I reduced the third data point; the sensitivity was worst of all. And then I looked at the data, and I felt
a wave of satori, nirvana, joy, peace, love, what everybody who
enters science is hoping to find. Because if you ran in the other
direction, the sensitivities, changing so rapidly over such a limited
range of longitude, would indeed go to zero. I was right.

And that turned out to be enough for Jerry to permit me to stay
for the rest of the week. Because this was a true and genuine
discovery, we had to follow it up.

Of course, I took advantage of that week to get as much material
as I could on what my colleagues were doing to go into my book
Colonies in Space. That is why when it finally reached print, it was
up to date. I also left that week with a draft of a paper for the
AIAA Journal[7] setting forth this new finding on how to use mass
drivers operationally. I knew that the whole problem was in
principle solved, and that we could get everything we wanted out of
this discovery.

When I got home, I found that I had a fellowship offer from
Germany. This gave me an opportunity to divorce my wife. Also, we
finally nailed down the contract for my book with Stackpole Books.
So all in the space of a single week, the week of July 19, 1976, a
number of major themes came together in my life, setting me on the
road to success in authorship, success as a scientist, and much more.
It was a wonderful week.

OTHER STUDIES

Now this was 1976. Out of that summer study in 1976 came
another volume published by the AIAA, volume 57 in their series
"Progress in Aeronautics and Astronautics."[8] It is, so far as I'm
aware, still available through the AIAA. It introduced technical
issues dealing with the mass driver and associated trajectories, as
well as some of the technical issues associated with early phases of
space colonization. My book came out in early 1977, only a few weeks
after Jerry O'Neill's book came out. Jerry did not love me for that.
He was especially angry when a review in the New York Times described
my book as "definitive." He, in fact, wrote me a letter saying, "How
dare you say such a thing?," as though I worked for the New York
Times, or wrote reviews for the New York Times. That's a very interesting letter he sent me.

There was another conference in 1977. I came back from Germany
to give a paper at the conference and my book was on sale. Some
people were able to buy it there. Perhaps some of you who are here
today bought it at that conference when it was available for the
first time. It was fun signing autographs left and right.

That summer there was another study.[9] The message I received was "Tom, you crashed the last summer study, but you're not going to crash this one!" I stayed in Germany that summer pursuing my own research program and carrying forward studies of these special trajectories which in due time led to a series of three full-length articles in the Journal of Spacecraft and Rockets.[10],[11],[12]

After 1977 things began to settle down. During the years 1974 through 1977 the ideas were sufficiently new and sufficiently fresh, there was enough new stuff to be learned and to be developed, that it was worth people's time to look into these matters. After 1977 things came somewhat to a halt. Since then, for the most part, people have been quoting and citing references from those earlier years.

There were a few other things that happened. For one thing, in 1979, there was a short conference at JPL, concerning non-terrestrial resources. I had the pleasure there of meeting David Ross, who shared with me a penchant for German songs and much else. We went out to lunch at Arby's and there were no other customers there, only the people who were serving the Arby's roast beef sandwiches and things like that. Dave and I proceeded to sing "Die Wacht am Rhein" and various other songs. Whereupon the girl behind the counter said to us in a very sweet way, "Would you perhaps care to go next door to McDonald's?" And that is known as the time that Dave and I got kicked out of Arby's.

But Dave and I, that summer, did a study along with another friend of mine, Eric Hannah, who had been an associate of O'Neill in earlier days and was now working out in Silicon Valley. We did a study on some further mass driver problems which led to two papers in the AIAA's Journal of Guidance and Control published in 1982.[13],[14] Dave and Eric, being denizens of Silicon Valley, eventually founded a company called Palantir which is now worth several million dollars. I, living 400 miles away, was not in on that and so I am not worth several million dollars, alas.

There was one other flurry of interest in 1980 when we had the second energy crisis, and power satellites once again elicited some very modest degree of interest. The Department of Energy hosted a conference on powersats in Lincoln, Nebraska. A nice central location, everybody could go there while paying the maximum airfare. Not like getting to the East Coast. That conference gave rise to a particularly thick document.[15] O'Neill had another conference in '81 which has led to some sort of proceedings volume. He had another conference in '83, still another in '85. But how often can you reheat the same soup; how often can you rewrite the same material?

So my view of the story of Space Colonization and the story of its literature is that between 1974 and 1977 a series of ideas elicited a certain amount of interest, even a certain amount of excitement. They gave rise to some literature, and certainly launched me on my career as a writer for which I'm very grateful. In fact, most of what I do these days, most of my friendships, associations, and activities in one way or another, can be traced back directly to space colonization. Let me give you an example.

MY SECOND BOOK: A VIEW OF LIFE IN SPACE

I've mentioned my first book Colonies in Space. There was also a second book which is not so widely known, alas, called Toward Distant Suns.[16] I happen to like Toward Distant Suns better than I like Colonies in Space. Colonies was largely a rewrite of material that was readily available, at least to me. For example, during May of 1976 I had about three chapters to write in the space of six days because of certain deadlines. I remember having a very vigorous battle with my wife at the time because I was busy trying to get the work done and she wanted me to do some babysitting. She didn't understand that it was significant. So I had three chapters to do in six days and the only way that you can do that is if you are basically taking material and rewriting it (which is another piece of information for any of you who wants to write about plagiarism as an important factor in my career).

That was Colonies. Distant Suns was different. Distant Suns was for the most part new. I really had to do original research and dig up original material. I couldn't just take some paper from a conference the preceding year and change the title to serve as a chapter. Well, I didn't do it like that, but it was almost the same. Let me give you an example of the kind of stuff I put into Distant Suns.[17] One of the things that I enjoyed in that book was addressing the question 'What sort of society are we likely to have in a space colony?' -- a question which I have reason to believe is of some interest to some of the participants here.

The point of view I tried to develop was that it would likely resemble the society of the Panama Canal Zone. The Panama Canal Zone is a place that is near and dear to my heart. I lived there for a number of years. I cherish a number of friendships from that time in my life. Now what is there about the Canal Zone that makes it relevant and pertinent as a model for a space colony? Well, how did the Canal Zone come into being? It came into being as the focus of a major national effort to direct the most advanced technology of the day, the day being 1910, upon a very restricted geographical area to pursue a long-held dream. It was a great national goal. The techniques at hand were steamshovels and cranes rather than mass drivers and whatever else. But the basic idea was the same, that of applying the best technology and the best resources to this difficult task.

And what came out of it after the canal was completed, after this great national effort had been brought in, under budget, I would add, and within less time than they had projected? What came of it was a collection of small, homogeneous communities which were then called the Canal Zone, and which I have reason to believe would resemble a space colony in a lot of ways.

What was it like down there? Well, they didn't have extremes of wealth or poverty. All of the people were government employees or their sons or daughters or spouses. The senior officials, who made a lot of money, lived 50 miles away from where I lived. The poor people, who came to clean our houses for $30 a month, lived across the border in Panama. The rich and the poor were out of sight. So we had a homogeneous collection of reasonably well educated Americans working at reasonably standard government jobs. There were no great opportunities for wealth or poverty or envy or, in fact, high achievement in our community.

Now there were some interesting differences between life in the Canal Zone and life as we know it here in the States. Take the system of politics for example. We were governed by a governor appointed from Washington, and he had his advisors around him. We had the right to vote. We just didn't have anyone to vote for. We could vote at the local level to elect a local town president who would now and then meet with the governor and tell him how we needed some new streetlights, but it's not as though we had city councils with taxing authority and municipal budgets. We didn't.

At the national level we had what might be called a gerrymander in reverse. If there had been enough of us down there, we would have had a congressional district, and we would have either dominated that district or at least had considerable clout within it. As it was, we voted by absentee ballot in the 435 districts from which we had originally come, a gerrymander in reverse, a nice way of diluting our voting power. As I say, we had the right to vote, just no one to vote for.

There were also some interesting elements of free enterprise. The government was the only organization authorized to offer employment. Even the lawyers were employees of the government, which meant that if you thought you were getting a raw deal you might have some problems finding counsel. The only way you could live in the Canal Zone was as an employee of the Federal Government. If you wanted to be self-employed or work for somebody else, you had to go live in Panama. You couldn't live in the Canal Zone. And there were some interesting procedures that they had in the days before court decisions and laws gave people stronger rights to their jobs. If you lost your job, you had to leave the Canal Zone in thirty days; you couldn't live there any longer. And it was not so hard to lose your job. As one governor said, "If you don't like it here, there's a boat leaving every Thursday."

For example, I know of some people who lost their jobs under some unpleasant circumstances. There was one fellow who shoplifted some cans out of the grocery, and he got sent back to the States pretty quickly. Another fellow bored some holes into the floor of his apartment because a pretty girl was living downstairs. He wanted to be able to get a look at her. When he was found out, he too was sent packing. The point being that here in the States if you or I did any of these things we might face probation; we might face a heavy fine; we might get jailed for a weekend for a first offense. But we would not lose our job or have our career damaged. We would maintain in the States a separation between our academic or professional activities and our life away from the job. In the Canal Zone, that distinction did not exist and if you were not keeping your nose clean, you would likely have to leave on the next Thursday boat or at least a boat traveling some Thursday later that month.

All of these things I feel are highly relevant to the way a space colony might be set up. Again, the people there might have the right to vote, just nobody to vote for. There might be extra-judicial procedures to send you packing if the powers that be didn't like you.

Another aspect of life down there had to do with freedom of the press. Did we have freedom of the press? Of course we did. We were Americans. Did we have accurate, independent, professionally reported local news? No, we did not. We had to rely on the party line published by the government or on the dubious charms of the Panamanian press which circulated English-language editions. If any of us were so bold as to seek to become journalists, we were likely to find ourselves offered employment by the government or told, 'What you're doing is incompatible with your work for the government and you better cut it out.' Remember that I said even the lawyers worked for the government.

I reported this scenario in Toward Distant Suns as one that could be highly relevant to the society of a space colony, a society of Americans expecting to live according to American customs, and with American rights, while depending for these rights on a powerful government that concentrates all responsibility and authority into its hands while conferring benefits upon its employees.

I'll tell one other thing that came out of Distant Suns. While I was at Stanford in the summer of '79, working with Dave Ross and Eric Hannah on mass-driver problems, I got word from my editor telling me that Distant Suns had gained success. In terms of royalties and sale of rights it was on a level with that of Colonies. In early September of 1979 I was at a party of the L5 Society, and I saw a friend of mine and said to him, 'Terry, guess what's happened!' I told him how we had just sold the rights to Toward Distant Suns. There was this incredible, beautiful, lovely redhead standing close by who worked for Terry in a lab at TRW. She wanted to meet this fellow who was an author. I was sitting on a couch with a plate of chicken; she came and sat down next to me. Her name was Angela. I took her out the next day. We've been living together for the last six years. She's made me very happy. In that respect and so many others, Space Colonization has been a very important influence in my life. With that, I thank you.

REFERENCES AND FOOTNOTES

1. Space Manufacturing Facilities (Space Colonies), Proceedings of the Princeton/AIAA/NASA Conference, May 7-9, 1975, (Including the Proceedings of the May, 1974 Princeton Conference on Space Colonization), edited by Jerry Grey. (American Institute of Aeronautics and Astronautics, Inc., 1290 Avenue of the Americas, New York, 1977).

2. *Space Settlements: A Design Study*, NASA SP-413, edited by Richard D. Johnson and Charles Holbrow, (U.S. Government Printing Office, Washington, D.C., 1977).
3. Gerard K. O'Neill, *The High Frontier*, (Morrow, New York, 1977).
4. Thomas A. Heppenheimer, *Colonies in Space*, (Stackpole Press, Harrisburg, PA, 1977).
5. T.A. Heppenheimer and Mark Hopkins, "Initial Space Colonization: Concepts and R & D Aims," Astronautics and Aeronautics, 14, 58-64, March, (1976).
6. Brian O'Leary, "Mining the Apollo and Amor Asteroids," Science, 197, 363-366, (July 22, 1977).
7. T.A. Heppenheimer and D. Kaplan, "Guidance and Trajectory Considerations in Lunar Mass Transport," AIAA Journal, 15, 518-525, (1977).
8. "Space Manufacturing from Nonterrestrial Materials," Progress in Aeronautics and Astronautics, vol. 57, edited by G. K. O'Neill, (AIAA, New York, 1977).
9. *Space Resources and Space Settlements*, NASA SP-428, edited by John Billingham, William Gilbreath, and Brian O'Leary, (U.S. Government Printing Office, Washington, D.C., 1979).
10. T.A. Heppenheimer, "Achromatic Trajectories and Lunar Material Transport for Space Colonization," J. Spacecraft and Rockets, 15, 176-183, (1978).
11. T.A. Heppenheimer, "A Mass-Catcher for Large-Scale Lunar Material Transport," J. Spacecraft and Rockets, 15, 242-249, (1978)
12. T.A. Heppenheimer, "Steps Toward Space Colonization: Colony Location and Transfer Trajectories," J. Spacecraft and Rockets, 15, 305-312, (1978).
13. T.A. Heppenheimer, D.J. Ross and E.C. Hannah, "Electrostatic Velocity Adjustment of Payloads Launched by Lunar Mass-Driver," J. Guidance, Control and Dynamics, 5, 200-209, (1982).
14. T.A. Heppenheimer, D.J. Ross and E.C. Hannah, "Precision Release and Aim of Payloads Launched by Lunar Mass-Driver," J. Guidance, Control and Dynamics, 5, 291-299, (1982). Also in *Space Manufacturing*, op. cit., pp. 191-201.
15. *The Final Proceedings of the Solar Power Satellite Program Review*, U.S. Department of Energy Conf-800491, 1980.
16. T.A. Heppenheimer, *Toward Distant Suns*, (Stackpole Books, Harrisburg, PA, 1979).
17. My books are now out of print. *Colonies in Space* may still be available in paperback; I'm not sure. I know that it is out of print at Stackpole. *Toward Distant Suns* never reached paperback. Fawcett bought the rights -- paid a very tidy sum -- and never published it. I now grant my leave and benediction to any of you who wish to copy from this material for your use in classrooms and whatever else.

PANEL DISCUSSION BY PARTICIPANTS
IN THE 1975 STANFORD/ASEE/NASA SUMMER STUDY
ON SPACE COLONIZATION

TEN YEARS OF PERSPECTIVE ON SPACE COLONIZATION

Introduction

Arthur Kantrowitz (chairman): The panel members are veterans of
the 1975 NASA-Stanford Summer Study for which I think we must be
endlessly grateful. That study recognized more fully the problems of
radiation which had been glossed over pretty briefly until then, and
they made an honest attempt at a realistic design of a space habitat.
Now as we heard very eloquently from Dr. Heppenheimer, the enthusiasm
for these things seems to have waned for the moment, and I would
think that it's an important exercise for us to discuss here what
will again make it wax. I'd like the panelists to start from the
audience's left side, and each, please, give your name and three
minutes of comment.

Statements by the Panelists

Holbrow: I'm Charles Holbrow, Colgate University. Arthur has
described enthusiasm for space colonization as waxing and waning.
I'm not very worried about that. I do believe that space colonization in some form is going to happen, because humanity can not resist
the temptation of a new frontier. Believing that, you don't have to
do anything very extraordinary to make it happen.
    I am very enthusiastic about the theme which underlies this
meeting. Regardless of whether space colonization happens or not, it
is an extremely exciting idea, and it does stimulate young minds, and
it does have a very fruitful role to play in the curriculums of any
number of disciplines. As academics we often function with a narrow
focus. On the whole, that's not a bad thing, but it is a good
experience occasionally to have something that stimulates different
perspectives in different ways. For the past day and a half we have
been accomplishing that end here. I hope that we can take what we've
done here back to the schools from which we came and inject this
particular kind of fizzy life into the curriculum.
    Sutton: I'm Gordon Sutton from the University of Massachusetts
at Amherst. I certainly agree with the point that Charles is making.
I would disagree on one minor point. During a time of waning of this

0094-243X/86/1480141-13$3.00 Copyright 1986 American Institute of Physics

kind of interest, we should recognize that this kind of project also stimulates old minds. And that's a good thing.

I was stimulated to get involved with the Summer Study to begin with by a continuing interest in and concern with sociological questions about the organizational requirements for human communities. How do human communities organize themselves? What are the basic elements that compose them? For example, it's interesting that ten years ago in the Summer Study, we started talking about a population size of 10,000. This was a fairly arbitrary choice at the time. And that number hasn't changed. People still refer to something like 10,000 as the basic unit of population. But there really isn't any logical foundation for the number. Sociological work is not advanced enough, unfortunately, to answer the question: What is the proper size of a colony. It is a question that I think is important.

I also want to mention that work in the Summer Study has led me into teaching in the area of technology and society. There are a number of us here in the room who have teaching interests in this area. I've become very interested in the problems of the identification and measurement of risk and uncertainty in society. I partly come to that from the organizational point of view. I commend to you a piece of literature as a teaching resource, Charles Perrow's book Normal Accidents, which has a number of case studies in it that I think might be of interest, including one dealing with the Apollo program.

Russell: I am Allan Russell, Hobart and William Smith Colleges. Fortunately for us -- unfortunately for space colonization -- we live in a democratic society. This is not a project for a democratic society. This is a project for a totalitarian society. And it's going to happen; it's going to take place. The question is who will do it. Will it be a totalitarian society that develops the space colonies, or will it be a democratic society that develops the space colonies? Will the solar system be populated primarily by totalitarianists, people with totalitarian interests, or will it be populated by people who believe in a free society? My hope is that we will be goaded into doing it, just as we were goaded into going to the Moon, because we refuse to allow a totalitarian society to be the one to call the shots.

Giesbrecht: I'm Martin Giesbrecht and I'm an economist from the Cincinnati area. One of the ways in which economists aggravate everyone, including themselves, is that when they apply economic analysis to any problem, even the most heroic, they drain all the passion out of it. This is a problem that can be overcome by applying ideas like space colonization to the teaching of the science. That's the big contribution that this program has made to me and to my students from my first contact with it in 1975 and in all the intervening years. I hope somehow through us space colonization can also make it into economics.

Heppenheimer: Tom Heppenheimer, author. As I review what's happened in the last ten years, I see a few narrow, limited areas in which we have made progress, even great progress, and a number of other highly important areas in which we have fallen back. I think the Strategic Defense Initiative, popularly known as Star Wars, is going to turn into a gold mine of technology that will apply to space colonization. We see coming out of Star Wars new approaches to the large-scale transport of materials, including major advances in the mass driver. Also we see major advances in computers in the general area of robotics. We know more than we did ten years ago. And we can anticipate such systems as the space spider which you will see described in Toward Distant Suns, which could serve for the construction of large pressure hulls and major developments in power satellites. That's the good news.

Now I'm going to tell you the bad news at somewhat greater length. First of all, ten years ago it was still possible to think of the Space Shuttle as the opening wedge for a new era of low-cost space transportation. Today we know that the Space Shuttle is no sort of an opening, nor is its successor likely to be anywhere close to what we would want. What's the answer? I don't know. The Defense Department is undertaking an initiative aimed at building an aircraft, a power rocket, that can fly from the runway to orbit and return and do so repeatedly, not several times a year but several times a day. Is that going to pan out? Come back in ten years. Maybe then we will have some hope. For the moment the prospects for low-cost space transport look farther away than they did in 1975.

Another important area is the processing of materials. We had no clear or good ideas of how to do the processing of materials in space in 1975, and we still don't. We still don't know how to build that steel plant on the Moon or that aluminum plant. Even though we have some interesting speculations, we have nothing that has been shown to be feasible.

Another overarching problem is that of economic justification and rationale. In 1975 it was possible, barely, to consider that the power satellite might be interesting and significant. Now we could say that the power satellite measured against the needs of America for energy is wrong on two counts. First, America's needs are in the realm of fossil fuel; we would like to have more oil. We are eventually going to have to turn to our coal. Well, we have four-hundred years of coal. For people who ask, 'What happens when the coal runs out?' my answer is, 'When Queen Elizabeth was preparing to fight the Spanish Armada, she was not framing policy initiatives that would reach fruition in the time of Margaret Thatcher. Four hundred years is a very long time.' Secondly, even if you grant that we do have a problem with fossil fuels because of oil and whatnot, and that we must develop electricity, if we need a lot more electricity we can either build more coal-fire plants or we could GO NUCLEAR! We could build lots of nuclear plants which is what France is doing. France takes its energy and electricity needs seriously and has been turning out plants like cookie-cutters. The power satellite isn't going to work on those two counts. Economically the power satellite isn't going to cut it, and technically we don't know how to do it. Maybe space tourism, maybe orbiting hotels in lower-Earth orbit are going to be an important entering wedge. That's an interesting speculation. But on the grounds of economics and technology we are farther away now than we were ten years ago.

Jebens: I am Harry Jebens, Marquip Corporation. Another area we are further behind in is life-support systems. We haven't put any money into that for years, and the best people are leaving the field as they retire.

Hill: I'm Pat Hill, an architect from Cal Poly, San Luis Obispo. I think I am the only person from the visual arts or design fields who came or even applied to join the 1975 Summer Study. That summer we recognized the complexity of the problem of dealing with a large number of people in an alien environment. We tried to address key issues, especially the need to provide diversity in order to allow a number of people to live in such an environment and have a sense of choice and a sense of control of their own destiny. Recognizing this need is a lesson I learned that summer. I hope that any future endeavor into space recognizes that such diversity has to be provided.

And certainly it would have to be recognized that on Earth, with the number of people we have and their diverse interests, these issues have to be addressed and conflicts somehow resolved before we can take on the effort of going into space. What came out of that Summer Study was only an idea and, for me, just an exercise. For the future we would have to allow all kinds of opinions, such as we've heard in the last day and a half, to surface; then eventually something may happen. What that might be no one can say.

The designs in the 1975 Summer Study were strictly a one-night sketch problem during the tenth week of the Study. Something was needed. I got my drawing instruments out and did an all-nighter. Unfortunately, that design probably has been looked at by many people out of context. It was never meant to be a definitive solution. The design was only one of many possibilities for an interior; I think an interior design for humans would have to come out of the needs and aspirations of the people and not from a preconceived solution. That's an issue that the future designers of any kind of space habitat would certainly have to face.

Brody: My name is Steve Brody and I'm with Intermetrics which is a NASA contractor. The perspectives I've had the last ten years are two sides of a coin. One side is a lot of change that I've seen. I was very fortunate to be one of the many hordes that helped develop the Space Shuttle, and I've seen that come to fruition as a significant engineering achievement. I have also seen things around the periphery changing the nature of space in the past ten years. There have developed commercial enterprises and a multiplicity of space-capable nations and entities rather than just a bipole of U.S. and Soviet interests in space.

We are on the verge, I think, of many significant advances in engineering and construction materials in space which will make the space environment more controlled. It will become a place still not totally familiar, but one where we don't fear an inability to grasp some new situation because we have so little experience with it.

But I see no changes taking place in a lot of the issues we raised here at the conference. We have rehashed a lot of things we discussed ten years ago in a similar way. I guess ten years from now we'll probably have the same discussions again.

Some of the questions about what the social environment of living in space might be like are being dealt with right now. I'm now involved with the Space Station plans to some extent, and a lot of talk takes place about how the inhabitants of the Space Station will function and interact. The model that seems to be coming to the surface is that of a space station which is a facility with operators providing a pseudo-commander type environment. You're going to have users of that facility who really don't need to interact too much with the operators unless there's a problem. Users will have to deal with an arrangement where a commander does take charge and say, "We have to get this done because of some life-threatening situation." But it will be a fairly benign and interactive mode between lots of folks.

One thing space colonization does for me is it allows me -- I'm sort of into trees -- it allows me to sit back and say, "Wait a second. What's the forest? Why am I spending my time working on space as my career?" It allows me to step back and look at the little steps we're taking now and where we're really heading. And if anything, I see space offering the opportunity for multi-national, multi-entity cooperation to work for human-kind and perhaps to tap a synergism of people and countries working together. Of course, it also offers the opportunity to fail in that enterprise. But I hope that Space Station and anything that follows these little steps we take, can be a test bed for our willingness to try to dilute the nationalism that's often created more problems than it has created solutions.

Lawrence Winkler: My name is Wink, I have several others. About ten years ago I had an incredible opportunity to participate, in a larger measure than I first thought, in determining certain physiological design criteria for the space colony. These were very, very rigorous criteria, and I have waited ten years to see some relaxation of them. I haven't seen any evidence at all in the last ten years in terms of research into certain of the areas.

But it's very funny whenever people come anywhere near this topic, there is an insane, magical feeling about it, and, yes, it's cross-disciplinary and everything else, but there's something about it that has a great deal of attraction not only in terms of its cross-disciplinary applications but just because it's a frontier of some kind, if I can quote that word from Hep and others.

The last five years I have been traveling around the world, because, I guess, I couldn't afford to wait for NASA. When I came back, numerous of my friends said, "Where are you going to go now? You've seen everything." And I said, "I'm going to go to Geneva and talk about the next step."

Rowland Richards: My name is Toby Richards and I'm at the University of Buffalo. Coming last, of course, I've very little to add. I think there was the feeling in 1975, in fact, we sort of jokingly referred to it, that space colonization might come in the year 2017, the 100th anniversary of the Russian Revolution. There is a question of will. It's very hard to marshal that in a democracy. But at the same time, something like space colonization as a first step beyond living in orbiting tin cans, which doesn't appeal to any of us, is needed as an answer to the dead-end type of NASA program which we've all experienced with Apollo.

Also, when the 1975 Summer Study was set up, all of us arrived excited. We had just finished reading the Club of Rome Report. It's very important to recognize that space colonization is an answer to the limits to growth. That's what originally stirred us all up. We had been sitting there pessimistic and then realized that we had done what Dr. Kantrowitz described: We had expanded the volume. That gave us tremendous drive. And it's still an answer. And I think it's the only place, it's the only way, that you can actually build a starship. You cannot build a starship on Earth. You have to build it in space. And as the first step for a starship it is essential. I'm as excited about it now in my relaxed way as I was then. I looked on it as a task for my children, and I still do. I think we all do.

Kantrowitz: We look at the problem too small. Too small, that is, if we look at space alone. We must look at the other things that have been happening, particularly in America in these years. This has been -- starting in the 60s -- what I like to call a time of timidity. This is a time when we became too timid to exploit nuclear energy, when we became too timid following thalidomide to permit at a reasonable rate the introduction of novelty in medicine. It has been a time when genetic engineering, which offers promise comparable to that of space, has been attacked -- so far not very successfully -- but attacked by the same forces that have slowed down the use of space. We must recognize that there has been a large-scale social change from optimism, for which America was always known, to pessimism in which we now lead the world. We are the most pessimistic nation on the face of the Earth. We swing. When we recover our optimism, all of the things of which we dream today will suddenly become possible. Now I'd like to get questions from the audience.

Questions from the Audience

Barry Turner: Following up your remarks, Professor Kantrowitz, I would like to raise an issue to put to the members of the panel. You observe the opposition of optimism and pessimism. Can I offer a substitute theme, a strong trend in the American culture as I see this from outside the society. This culture has always had a very strong belief in an instrumental, interventionist relationship with the environment. You fix things in the environment; you did things in it. You didn't look at the ways of the native Americans who lived here before colonization.

The question I really want to put to members of the group is whether they see this interventionist approach as one which they would see positively continuing in the space-colony setting or whether they see it as a form of escapism which seeks a technological fix, a solution which diverts attention from problems that are difficult to solve here but which will pursue you even when you have a technological fix? Is there any sense in which members of the group would see the movement into space as a violation of space rather than as an isolated effort to develop a settlement with a Howard Johnson's?

<u>Giesbrecht</u>: I think that both of those views are compatible. That's the extraordinary thing. Speaking as an economist and one of the reasons why I'm here and involved in the problem is that you know and I know and we all know that when we finally do make our steps into space, they will be businesslike incremental steps. Little additions here and there which will at some time ultimately look like space colonization although it won't be called that at first. That's very drained of passion and many of you in the audience and many of us, myself up here at the table, find that not particularly an attractive view, but it will be the way it's done. Nevertheless, with that incremental movement into space which is so businesslike -- Howard Johnson's and McDonalds -- there must also be a kind of a fire that makes meaning out of this. It seems to me that therein lies the balance of everything we do, which is after all very ordinary and everyday if you look at it from the economic standpoint, but begins to take fire and shimmer in our imaginations when we finally view it in a whole picture and see what it all comes to. That's life isn't it?   And I think in that regard, space exploration is the same thing. Let's not lose sight of that wonderful passion that lends so much meaning to the everyday steps that we take. I have no response to your last question, by the way.

<u>Kantrowitz</u>: I think the issue here is power. When you speak of violating space, if for you that is immoral, then don't do it. If for me the opportunity to provide living space for humanity is very moral, then I'll do it. And let's be friends.

<u>Hill</u>: I sort of agree with that. When a crisis arises the ends justifies the means; that is obviously what happened when they dropped the atom bomb. I think if a sufficient crisis did arise in the future of the Earth, there certainly would be a completely different attitude and environment that would probably motivate or justify an impassioned effort. Right now we're complacent as a society and comfortable (or uncomfortable to some extent) with the world's economy. We seem to have sufficient oil and resources to keep us going for another few years. But I imagine if something happens to create a conflict, then space colonization might be seen as an alternative. Major changes have always occurred when that kind of crisis has arisen. I can see it happening again, but when and under what circumstances I don't know.

<u>Heppenheimer</u>: I want to talk about this idea of violating space. Unfortunately, we have heard far too much of this type of view in recent years. It is part and parcel of what my distinguished colleague has referred to as the 'pessimism of our time.' Much of this pessimism, I fear, stems from a view of man as a pitiful, weak creature who must live huddled around his hearth fire never dreaming to look at the stars above. I say NO! Mankind is proud. Mankind is defiant. Mankind will move forward and will advance and will do those things which are feasible, which we wish to pursue.

<u>Turner</u>: I'd like to get back to space violation. I have in mind the image of those traditional cultures of the past in which if

you were sacrificing an animal you were then mindful of the spirit of that animal. If you were mining a mountain, you would be sure not to disturb the spirit of that mountain. It seems to me that a lot of the despoilation that was associated with the development of America is actually the absence of that kind of reciprocity with the environment. I was inquiring whether there might be some sense of such reciprocity, an idea that space was not put there merely as a resource for our exploitation. Is there some reciprocity between this resource and those who are entering into it?

Russell: To call this effort a frontier is more accurate than perhaps we think. The idea of living in harmony in an environment is an equilibrium idea. The idea of a frontier is not an equilibrium idea. There's a sense of threshhold, a sense of novelty, of unknown, of danger to survival. And under those circumstances, the instrumental approach to existence, it seems to me, is not only entirely appropriate but necessary. The equilibrium, the harmony, comes afterward. It's entirely appropriate on the Earth, in the environments that we have developed and populated, to try to move toward that harmonious equilibrium. I think it's quite inappropriate to see that harmonious equilibrium as an early objective at a frontier such as space.

Winkler: I'm not sure where in the upper atmosphere our philosophical view should change. But I would make one small point. If building an entire eco-system from scratch and putting it in a hostile environment like space doesn't teach us anything about our own eco-systems, I'd be very surprised.

Voice A: I have a question for Tom. In Toward Distant Suns you talk about the economics of satellite solar-power plants. What do you think of the prospects now? Can they be built more cheaply in space?

Heppenheimer: The experience of utility executives has been that in recent years they don't even know what a coal-fire plant costs. They can't realistically project costs of a coal-fire plant or anticipate meeting or beating those costs seven or ten or twelve years down the road.

The situation for nuclear is at least as bad, possibly worse. Much worse. Not long ago I was talking with a utility executive who told me that he had costed out Diablo Canyon at $300 million in the early 1970s or late 60s. At the time we talked the cost was $2.4 billion and still rising. He said, "Thank God I haven't had anything to do with Diablo Canyon in at least the last seven years."

If this is the case for energy systems that for decades have been built, put into operation, and used, what can we say then about cost estimates of technologies that do not now exist and which cannot reasonably be projected and with which we have no experience? We may speak of cost goals, but that is a far cry from realistic cost estimates.

Voice A: What do you mean by cost goals?

Heppenheimer: We don't know how to build satellite solar power plants, and until we do and until we gain operating experience with them, what are put forth as cost estimates can only be regarded as cost goals.

Kantrowitz: I'd like to say in response to what Tom said that I heartily disagree with him. The cost of nuclear power is exactly as projected if you add the cost of timidity to it. The fearmongers in our society who dominate today, have added the extra cost, have made nuclear power impossible in this country. Fortunately, they don't rule the whole world yet.

Jebens: At one time, to get a free dinner, I wrote a paper on "Power in Space." The numbers that I dragged out I have mentioned previously, but I think your question was 'Could you do it cheaper from space?' I'm not sure what the cost is to Earth if you had 10,000 people in space, completely self-sufficient, not using any of Earth's resources, building power stations. What is the cost to Earth? The numbers we had were 20 percent of the launch cost.

Giesbrecht: Let me just substantiate what Harry said there. When we are talking about economic costs here on Earth, it's difficult to know exactly what anyone means, because they frequently think of investments as costs when they're really not costs, they're investments of capital put into place. When we're talking about costs of things being done in outer space by people who live in outer space, then earthbound cost estimates really are utterly inequitable. The whole question then of who benefits from these things, who makes the profit from the sale of the electricity, is also up in the air. You have to recognize that when you buy electricity from a space colony that happens to own a couple of satellites out in space, you won't be paying ConEdison, you'll be paying those people out there.

We've heard a lot about the political science of what's going to go on, and I quite disagree. Those things won't be colonies for long at all. They're going to be independent groups of people who'll have some sort of relationship to their original country, linguistic probably, but it's very unlikely that the trade we'll have with space habitats will resemble anything other than international trade. We'll be buying electricity from people who are outside of our nation as if we were buying it from Canada. So I don't think it's useful to talk too much about costs of space. The concept is too ambiguous.

Heppenheimer: It is very ambiguous. We don't know how to build powersats. For that matter, we don't know how to build what could well be the competition for powersats. We don't know how to build fusion power plants, for example. We don't have nearly as strong a handle as we would like on the next generation of nuclear plants. There is an awful lot to learn and an awful lot of competitors with the power-satellites concept, many of which are readily extrapolated from present-day technologies. That's why I'm very reluctant to cast aside what we know how to do and with which we have problems that are known in favor of what we don't know how to do and where we haven't run across the problems yet.

I would also like to talk about the matter of the frontier and about the ecology that exists in equilibrium. For the human race, the situation of equilibrium is one in which one square mile of hunting range supports one hunter, which means that the gross human population world-wide can only be of the order of a hundred million. That is an equilibrium situation. We moved beyond that equilibrium, that ecological balance, during the neolithic age. And today our numbers are not in tens of millions but instead in billions. There is no going back. The people who speak of ecological harmony and equilibrium, in fact, are speaking of death sentences for 97 percent of humanity. Considering that their close ideological allies did just that in Cambodia, I'm not surprised. But I would hope that by promoting disequilibrium, that we would do better for the whole of mankind.

Barbara Eckman: However, concern for equilibrium did bring back the concept of harmony we spoke of earlier. It seems to me that you have a penetrating fantasy, as a psychologist would say, about what you're up to, exploring a frontier, pushing into the envelope, whatever you want to call it. You can penetrate without breaking.

I'm using the word 'fantasy' purposely because fantasizing is what's normally done in these circumstances. I'm saying that's fine. A penetrating fantasy is fine, as long as it doesn't become a rape. I don't hear most of you talking about wanting to rape the environment. To try to draw this analogy (or this fantasy) out a little more: struggle with the hostile environment is literally true; there is no error, but you don't want to take that too seriously and see it as an ending. That's when you start breaking stuff. Harmony for me is a kind of love -- here, a love of space, and you treat your lover well. You don't want to treat space as your rape victim but as your lover.

Heppenheimer: I would hope that we would all see penetrating insights rather than penetrating fantasies.

Winkler: As a matter of semantics maybe it's better to say it's an inhospitable environment rather than a hostile one. It is a good point. The absence of air up there is not the only thing that is inhospitable; there is the radiation, etc. And rather than a male fantasy of penetrating or anything else, it could be seem as a female fantasy in the sense that you are building a womb, you are constructing a womb in space. Thus, you can use the Jungian approach to give quite different views of entering space, but, essentially, I don't consider it rape.

Voice C: I'd like to make two points about the pessimist-optimist dichotomy. Whatever your view of whether the public should be more optimistic about technology, I think it's crucial that we recognize that the public mood has causes. There is an explanation of why the public mood is as it is. And I would contend that one of the principal explanations for technological pessimism, to the extent that it exists, is that the negative impact of technology has been all too evident. To say you want to boost the technological optimism which I think is probably everyone's working goal, is not something like saying you want to boost team spirit. You have go out and address those negative factors and explain why the negative technological pessimism exists.

In a sense it is almost beside the point to talk about technological pessimism as opposed to technological optimism. Technology is just the instrument, and the optimism or pessimism probably ultimately relates to the particular problems that technology would be brought to bear on or alleviate in some way. Allan commented about the problems in a democracy of promoting programs like this.

151

People who believe space colonization is important have to convince other people that this is the way to address human problems. A way to instill optimism in the human population is to say "Here are these problems; we have the means available to us to solve them; and we can do it." If space colonization is not seen by a large part of the population as something that they would support, it's because they don't see it as addressing all the problems

Kantrowitz: You don't see population pressure as a problem. Is that what you're saying?

Voice C: I'm talking about a way to perceive the environment.

Kantrowitz: But you said people do not perceive population pressure as a problem? I would say the world perceives that.

Voice C: Yes, I agree. The connection has been made by those who are in favor of space colonization. You haven't yet made the connection in the public mind between real problems and space colonization as a solution.

Kantrowitz: Well, technological pessimism means you should not try to do anything that hasn't been done before. There are the problems, as you can see, and there is the solution. But we are not developing the technology to get people there, because of the pessimistic perception of technology.

Sutton: I'd like to comment on that point and relate it to something that Hep said earlier, where the two of you disagreed. I think I am inclined to go with Hep if I understand him correctly. One of the lessons that we can learn from the Summer Study, looking back over ten years, is that our conception of off-the-shelf technology for designing the space colony failed on a very important ground. That is, engineering practice dictates that a great deal more discovery is required in order to build something which we claimed could be done with off-the-shelf technology. I think Hep's remarks about what it is that we don't know about building manufacturing facilities in space, or about building a lot of other things in space, in part don't have to do with the concepts of how to do these things, but with the practical, detailed physical lessons that have to be learned in order to move effectively in those directions.

Let me give you a case in point. The University of Massachusetts at Amherst has a library that is about 25 or 26 stories high. The library was abandoned in some sense about 10 years ago because bricks began falling off it owing to a very serious error in how the building was put together. There is still work going on to try to fix it, but it has been 10 years. We do not have the technology to solve what seem to be pedestrian issues of how to build power plants, of how to build structures, how to do a lot of things. The real-world extant technology is not as real as I thought, as I think we all thought, it would be.

Kantrowitz: I would assert that just as technological optimism can do most anything, technological pessimism can destroy, can create these instances of failure to which you refer. Twenty years ago we knew how to make nuclear power plants. Today we don't. And they did not hurt anybody; they were safe. Today we don't. That is the difference between optimism and pessimism.

Now the causes of technological pessimism to which you referred, if I may get back to that, what are they? We lost, the world lost, several thousand malformed children to thalidomide. That was a tragedy. The answer that technological pessimism gave was "Abolish new drugs." And effectively, if you want the latest treatment today, in most any disease, you must go abroad. This is the United States today; we're the leader in technological pessimism.

Heppenheimer: I want to comment on this famous cliche which we hear so often, "Oh, technology has got such horrible side effects. Technology has been so destructive." Tell me about the destructive effect of technology. When I say "tell me," I mean tell me in terms of people killed or physically injured. Do we have steamboat explosions any longer? No we don't. Do we have railroad boilers exploding any longer? No we don't. Do we have badly designed buildings in which huge numbers of people are lost because of fires as in the horrible Triangle Shirtwaist Factory disaster of 1911? No we don't. Do we send passenger liners full tilt through ice fields without full complement of life boats? No we don't. And do we have diseases such as the influenza epidemic of 1920 which carried off huge numbers of people? Well, who knows how AIDS is going to develop, but at least as of a couple of years ago I could have said "No we don't." We have had enormous advances in all of these areas where there were grave threats to human life and limb. What do we have in place of these crude and genuine threats? We have had people, mostly academics, mostly liberal-arts types, who were envious of the attention paid to their engineering and physicist colleagues, who proceeded to invent, largely out of old cloth, terrible disasters due to nuclear energy. Terrible threats due to waste which you know we could easily dump in the ocean and get rid of it. I think that all of this stuff about the threat of technology, at least 95 percent of it, is nothing more than a put-up job by envious people who resent the attention paid to those who can do rather than merely talk.

Kantrowitz: I want someone who will respond to Tom. Now hit him hard.

Voice D: How does the public perceive technology? You say it's a problem of technological pessimism. If you want to address that then you're going to have to address the causes. You can't go around name-calling and saying, "It's the liberal-arts types!" That's not going to solve the problem.

My specific example is the environment. This has come very much to our attention. We did things 10, 15, 20 years ago which now turn out to be carcinogenic. Now people who were involved are getting cancer. People think we have to be careful about technology because we don't know what will be created in 10, 20, 30 years from what we are doing now. That is the attitude and that is what you're going to have to address if you want to address technological pessimism.

Kantrowitz: It's an interesting thing that you don't make the charge yourself. You just say the charge exists in the public consciousness. The fact that it is completely phony, with the single exception of smoking, the fact that it is completely phony, you don't address. It is phony.

Heppenheimer: Your point is well taken, Arthur, and I will ignore it. As long as the public wins enthusiastic support from the government of the United States and the tobacco lobby continues to puff away, I will not take seriously talk of a threat of cancer. Until they outlaw tobacco, I will not consider that this nation has any serious interest in cancer. But I will tell you where all this talk about carcinogens comes from, where all this talk about carcinogens arrives from, it comes from advances in the sensitivity of measuring instruments. Twenty and thirty years ago we often could not do better than measure parts per million. Today we can often measure parts per trillion, which is a million times more sensitive. And so then, if you are finding some chemical which exists in the environment at that level, you have grist for a scare headline which would not have even been physically possible twenty years ago. But that does not mean you are affecting human health.

Voice E: I'd like to respond to Dr. Heppenheimer's comment about the distinction between those who act and those who talk. I feel personally that unconsidered and irresponsible action is more to be abhorred than irresponsible talk. And I think we've heard quite a bit of that here this morning.

Heppenheimer: I think we have heard a great deal of irresponsible talk in the last twenty years. That is what we technological optimists fight against.

Kantrowitz: I would assert that talk is a form of action and they are not to be distinguished.

Russell: Mr. Chairman, I would like to come back to the subject of this conference and say that "Space Colonization" was only the head title of it. The secondary title of it was "Technology and the Liberal Arts." We have now, at the very end of this conference, come to the problem: how do we foster communication between people familiar with each? The goal of this conference has been to elicit such communication with the hope that in the long run there can be cooperation. Sometimes at the beginning of the communication things go difficultly. There is a lot of stress; people don't understand each other; and they talk past each other. I think we're still at that stage. I hope that before too many more of the conferences with objectives such as this one go by, that we will get to the point where we begin to communicate more effectively and share the fruits of such communication with our students.

Holbrow: I would like to make a closing remark. This conference was opened by the Provost of Hobart and William Smith Colleges, and he mentioned that the conference has been jointly sponsored with Colgate University. On behalf of Colgate let me express gratitude to those people who came and participated. It has been an extremely interesting conference and certainly a stimulating one. We who organized it knew certain participants would be very stimulating indeed. They have not disappointed us. I thank you all.

AIP Conference Proceedings

		L.C. Number	ISBN
No. 1	Feedback and Dynamic Control of Plasmas – 1970	70-141596	0-88318-100-2
No. 2	Particles and Fields – 1971 (Rochester)	71-184662	0-88318-101-0
No. 3	Thermal Expansion – 1971 (Corning)	72-76970	0-88318-102-9
No. 4	Superconductivity in d- and f-Band Metals (Rochester, 1971)	74-18879	0-88318-103-7
No. 5	Magnetism and Magnetic Materials – 1971 (2 parts) (Chicago)	59-2468	0-88318-104-5
No. 6	Particle Physics (Irvine, 1971)	72-81239	0-88318-105-3
No. 7	Exploring the History of Nuclear Physics – 1972	72-81883	0-88318-106-1
No. 8	Experimental Meson Spectroscopy –1972	72-88226	0-88318-107-X
No. 9	Cyclotrons – 1972 (Vancouver)	72-92798	0-88318-108-8
No. 10	Magnetism and Magnetic Materials – 1972	72-623469	0-88318-109-6
No. 11	Transport Phenomena – 1973 (Brown University Conference)	73-80682	0-88318-110-X
No. 12	Experiments on High Energy Particle Collisions – 1973 (Vanderbilt Conference)	73-81705	0-88318-111-8
No. 13	π-π Scattering – 1973 (Tallahassee Conference)	73-81704	0-88318-112-6
No. 14	Particles and Fields – 1973 (APS/DPF Berkeley)	73-91923	0-88318-113-4
No. 15	High Energy Collisions – 1973 (Stony Brook)	73-92324	0-88318-114-2
No. 16	Causality and Physical Theories (Wayne State University, 1973)	73-93420	0-88318-115-0
No. 17	Thermal Expansion – 1973 (Lake of the Ozarks)	73-94415	0-88318-116-9
No. 18	Magnetism and Magnetic Materials – 1973 (2 parts) (Boston)	59-2468	0-88318-117-7
No. 19	Physics and the Energy Problem – 1974 (APS Chicago)	73-94416	0-88318-118-5
No. 20	Tetrahedrally Bonded Amorphous Semiconductors (Yorktown Heights, 1974)	74-80145	0-88318-119-3
No. 21	Experimental Meson Spectroscopy – 1974 (Boston)	74-82628	0-88318-120-7
No. 22	Neutrinos – 1974 (Philadelphia)	74-82413	0-88318-121-5
No. 23	Particles and Fields – 1974 (APS/DPF Williamsburg)	74-27575	0-88318-122-3
No. 24	Magnetism and Magnetic Materials – 1974 (20th Annual Conference, San Francisco)	75-2647	0-88318-123-1